Electronic Access Control

Electronic Access Control

Gerard Honey

Newnes
OXFORD AUCKLAND BOSTON JOHANNESBURG MELBOURNE NEW DELHI

Newnes
An imprint of Butterworth-Heinemann
Linacre House, Jordan Hill, Oxford OX2 8DP
225 Wildwood Avenue, Woburn, MA 01801-2041
A division of Reed Educational and Professional Publishing Ltd

A member of the Reed Elsevier plc group

First published 2000

British Library Cataloguing in Publication Data
A catalogue record for this book in available from the British Library

Library of Congress Cataloguing in Publication Data
A catalogue record for this book is available from the Library of Congress

ISBN 0 7506 4473 7

Composition by Genesis Typesetting, Laser Quay, Rochester, Kent
Printed and bound in Great Britain by Biddles Ltd, www.biddles.co.uk

Contents

Foreword

In recent years the security industry has made considerable progress in the development of educational systems and qualifications frameworks in support of a rapidly changing industry. The SITO/City & Guilds 1851 examination has been specifically designed to help the industry develop the pool of technicians required it it is to meet the demands of its customers and respond to technological and operational changes in the marketplace.

SITO is delighted to endorse this book as a supporting text for the SITO/City & Guilds 1851 syllabus.

I am sure that you will find this book of considerable assistance to your understanding of the access control industry and access control engineering. I wish those of you who proceed to take the SITO/City & Guilds examination or NVQs every success in your endeavours.

Raymond Clarke
Executive Director
SITO

The National Training Organisation
for the Secure Environment

About SITO

SITO Ltd is the world leader in developing accredited qualifications and training packages for the security industry. Formed in 1990, SITO is the UK's National Training Organisation (NTO) for the secure environment. It is instrumental in developing the skills of security professionals throughout the world. Working alongside international governments and other training organisations, SITO is establishing the highest possible standards in all parts of the security industry worldwide. Sectors covered include manned security services, retail and leisure security, parking, fire alarms, intruder alarm encgineering, CCTV/access control, investigation, aviation and transporting property under guard.

SITO's distance learning course in Access Control Engineering, a pack of seven units designed to introduce candidates to the terms and equipment used in modern access control systems, enables candidates to make informed decisions regarding the choice of equipment when proposed access control measures to meet defined security risks are required.

The course is designed to assist those studying the SITO/City & Guilds 1851 scheme and meets the underpinning knowledge requirements of the National and Scottish Vocational Qualifications (S/NVQ) for specifying, installing and maintaining access control systems at Levels 2 and 3.

For further information about SITO please contact Colin Reed at SITO on 01905 2000 4.

Acknowledgements

The author would like to thank all those who provided information for this book, and in particular:

Abloy Security Ltd
Bewater Cotag Ltd
Castle Care–Tech Ltd
Ellard Rib (UK) Ltd
Innovative Electronic Technology Ltd
National Approvals Council for Security Systems (NACOSS)
Paxton Access Ltd
Relcross Ltd
The Security Industry Training Organisation (SITO)
T.K. Consultants

Preface

Access control systems are becoming increasingly prominent in the electronic security industry in the management of entrances and exits and the monitoring and screening of personnel within defined areas. Such systems extend to the collection of data, the movement patterns of users and the integration with other security functions and building management services.

This book follows the Module for Access Control Systems covered by the syllabus of the City and Guilds/Security Industry Training Organization (SITO) 1851 award 'Knowledge of Security and Emergency Alarm Systems'. It contains the essential content of the course by following the units and will assist those who wish to study the requirements for specifying, installing and maintaining access control systems at Levels 2 and 3. While it endorses this course it extends its scope so that it can equally be used by all personnel with an interest in electronic access control systems on an international scale.

The chapters of the book have been introduced as separate units that form the framework of the access control industry. They will enable the reader to:

- Realize the terms and equipment used in modern electronic access control systems.
- Identify the key components of electronic access control systems to include the physical locking devices.
- Understand the main security applications of access control systems.
- Recognize the principles of survey, installing and maintaining these systems.
- Audit installed systems and diagnose and rectify faults.
- Appreciate the integration with other security systems and building management services.

1 Perimeter protection

This first chapter is intended to introduce the basic concept of electronic access control and to describe the main elements of such a system. It then details the barriers used in external protection with the means of operation of the different types and gives an insight into their rank in terms of physical strength and operating speed. Knowledge can then be gained into what perimeter protection factors should be considered for the different situations and how barriers can be designed to satisfy any specific application.

Consideration must also be given to safety aspects and the emergency exiting of sites.

1.1 Overview of access control

Access control in real terms is a huge subject because it can be extended to embrace all electronic security techniques by integrating with the other technologies, particularly intruder and CCTV.

The most basic form of electronic access control is a very simple domestic system intended to protect a locked but low security door so that a visitor can be interrogated by a telephone at a local point by means of a simple intercom. At the further end of the electronic access control scale is the intelligent computer-based barrier and multi-door high security system with remote surveillance and multiple users together with extra facilities for visitors. The former system may, in some parts of the industry, be classed as a door entry kit yet it still forms a part of access control engineering. The latter system will involve the use of many readers, by which personnel are checked as to their right to enter into a prescribed area with all transactions being logged on a database. Included in this category can be security perimeter fencing or walls with electronic detection that can provide electrical outputs to carry out surveillance practices with recording facilities over extended periods of time.

It follows that the electronic access control engineer must carry out diverse activities including the installation of mechanical locks and barriers and yet still be capable of programming systems either at a PC or at the system controller's integrated software.

An aspect that must be fulfilled with all systems is the installation of a locking device. All of these locking devices will have a capability to be unlocked electrically and must be monitored for their status together with the position of the barrier or door that they secure.

In the main, electrically operated release latch strikes and locks are easily installed in place of standard units to provide remote controlled access for a door. There are nevertheless considerations with respect to circuitry because of the various units due to the difference in sizes and holding forces. In addition certain doors need to be locked by electromagnets and the electrical rating and inherent technology of these devices must be considered in their own right since they often require unique power supplies capable of handling the greater electrical load they draw from the system.

Classifications are often made with respect to high security but with locks this is difficult to define and the term is regularly used arbitrarily by manufacturers to promote their goods. It is possible to argue that a lock advertised as high security should have features that offer more than ordinary resistance to picking, impressioning, drilling or wrenching. The more difficult for an unauthorized person to have a duplicate key cut it can be said the more security the lock provides.

Barriers form a separate subject and their security role is somewhat different in that they offer little resistance to determined human intrusion but are primarily intended to control vehicular traffic. Although any barrier can in theory be automated, the electrical requirements for this function vary from low voltage drives to mains power multiphase motors. There may also be certain heavy constructional building foundation tasks to complete for high security barriers and this demands specialist attention. Nevertheless the electronics engineer must, as a minimum, appreciate the role of other contractors in undertaking this work. In addition to this is a need to install sensing devices to ensure that the barrier cannot inflict damage on obstructions that could be positioned within its movement path.

There are more specialist forms of barriers and doors used in some applications but the technologies of driving and controlling these can easily be understood if the techniques used by the standard versions are appreciated. With all of these devices comes a need to install cabling and this must of itself be protected physically and be of the correct type to carry appropriate electrical loads and to resist attenuation of signals and corruption of data.

Allied to the larger system with barriers is the cabling for CCTV monitoring since barriers provide little resistance to human intrusion there may be an argument to employ beam interruption techniques to stop unauthorized persons from circumnavigating official access routes. This is a need if there is little security personnel presence to physically protect a sensitive site.

It follows that the perimeter protection which may be as basic as a low security door still requires electrical connections to qualify as an electronic access control system even though it will probably have a manual override in the form of a key or mechanically applied code. For this reason there is a requirement to understand the dangers of working with mains electrical cabling and the regulations as they apply. In addition many buildings have multiphase supplies and the means by which they are derived should at the least be understood even though there is no practical need for the access control engineer to work with them unless heavy duty motors are to be used to power barriers.

Electronic access control is a growing market and there are many kits available to satisfy many applications. In other instances separate components to make up a system can be purchased from different manufacturers if this enables a more effective solution to be realized. It is important however to survey for the future but not to lose sight of existing applications that have stood the test of time. This is apparent in mid-size applications such as buildings that house a number of apartments but have an external communal door which can be interrogated from all of the apartments and enable a visitor to be granted access by electrically unlocking the door from a telephone in the apartment. Authorized persons can open the communal door by use of their own key or by addressing the reader to prove their identity. These systems are common and historically have performed well. They are inherently reliable and feature standard wiring techniques. They also have features allowing trade persons to attend the building within prescribed hours and can come complete with duress outputs to signal alarms if residents are threatened. Such premises can now be upgraded to incorporate camera technology with no changes needing to be made to the cabling that will not be accessible in many properties.

It remains to say that in practice the surveyor can use existing systems as a benchmark but must be alert as to future trends and the need for upgraded security levels.

The cabling used throughout the range of electronic access control systems differs somewhat because of the diverse range of the products and networks so it is difficult to show exact detail but case examples have been included in the particular sections.

It is important to know that although the most very basic door entry kit may be seen as a stand-alone technology it is essential that we realize that the future is one of integrated control with other security systems and perhaps also building management.

We are now seeing integrated systems being introduced which themselves are in kit form. These are designed to meet the requirements of BS EN 50131, the new harmonized European standard for security systems, yet still have the flexibility to be used in a traditional British Standard sense such as BS 4737 covering intruder alarms.

These integrated kits are a modular system so that any specification can be built up from a basic number of different items to minimize stock holding. They can be wired as iD or using end of line resistors (EOL) to cover access control readers with full reporting, all standard intruder attributes/zones and to include fire and gas detection with personal attack outputs. Additional applications include call with mimic display, CCTV switching, lighting control, hold-up alarm and deterrent warning. Communications that are inbuilt within the package accept plug-on devices to cover all of the new generation and reporting techniques including digital communicator/STU/dualcom and digimodems for uploading/downloading and point ID reporting.

In their own right access control systems include an array of techniques to check the credentials of personnel and vehicles wishing to gain entry and exit from a specified area. The controllers and associated hardware may check for a user entering an authorized number code or presenting tags and cards to the system reader. Alternatively the verification of personal by biometric traits can be sought before the system perimeter locking will release. Dual techniques can be applied to increase the level of security so different credentials must be presented simultaneously before the system will accept an individual.

Using computer databases, all information can be logged so that it is always known the exact number of personnel and vehicles on a site at any material time. This can equally be used to plan working schedules and related salaries for employees depending on their times within the site. In addition the various grades of personnel can be ascribed different levels and times of entry within a site and to particular areas only so that all points are closely controlled.

The extent to which the system can be used and the complexity of its software is very much related to the equipment. This primary equipment has an influence on the cabling that must be employed but this will differ between the components used, the length of cable runs and the loading of the current consuming devices. In certain circumstances there will be a need to incorporate additional power supplies and in every case attention must be paid to suppressing electrical interference which can corrupt data. It is always possible to obtain from the manufacturer of the primary goods the cabling that is, in the main, used to terminate their products based on field experience but most cabling techniques do follow traditional patterns. The cabling must of itself be mechanically protected to withstand the anticipated degree of abuse and to achieve the correct degree of aesthetics. Consideration must equally be applied to environmental conditions not only for the cabling but for all of the system components so they are fit for their purpose in the environment in which they are sited.

Throughout the various chapters examples are offered so the installer can understand the options that are available.

Important if we are to survey for the future is the need to recognize the benefits of remote signalling so that a system can be interrogated and its parameters amended from a distant point that in theory can be almost anywhere in the developed world. This is achieved via a PC interconnection and modern communication techniques via hard wired telephone networks or radio links.

This brings us to the subject of handing over the system to the client to a clear company policy and plan. Maintenance, both preventative and corrective, must follow recognized practices and be endorsed by the governing response organizations in the industry. This all comes at a time when the security industry as a whole is being charged with maintaining and improving standards and this must be done on an international basis with harmonization being introduced wherever possible.

We can now summarize to say that the purpose of an access control system is to restrict the access of unauthorized persons to an area while facilitating the entry of authorized persons. It must include a method of operation to ensure that exit routes are not restricted and can easily be vacated.

In practice an electronic access control system must provide the correct combination level of access and control since there are two basic rules:

- The easier it is to gain access then the more likely the system is to grant access to unauthorized persons.
- An increase in the security of the system leads to a more likely denial of access to authorized persons.

Electronic access control (EAC) is a system in which information is collected and then analysed electronically with perhaps computer input so that instructions can be issued to other system components that provide locking such as barriers, electric locks or electromagnetic holding devices. It is these that provide the physical means of restricting access or granting entry beyond a prescribed point.

Access control engineering actually forms part of a much greater security system industry, extending through intruder, CCTV, lighting and fire so outputs can be taken from one system to activate another.

Building management can also be integrated with security concepts to form intelligent systems and remote signalling can provide data and information transmissions to central stations or other monitoring area to verify an activity. Enhanced communications then come about with tailored networks.

In the first instance however we should concentrate on the external physical protection devices and hardware.

1.2 External perimeter protection techniques

Access point hardware

The technique of protecting the perimeter at the access point is achieved by hardware. Hardware is the components that enable the access to be effective through a designated point of entry. The hardware that performs the physical function of allowing or denying access can include lockable/ movable barriers, turnstiles, gates and doors. In common use is the door complete with a lock of which there are many combinations. These are selected to suit the desired site security. At times these may need to be complemented with door open sensors to monitor the condition of the door and electrical or mechanical door closers to ensure that the door relocks following access through it.

There may also need to be protection afforded to tail gating by which an unauthorized person follows behind an authorized entrant and attempts to gain entry by close contact with the former person. The protection against tail gating can be afforded by turnstiles or 'air lock' door techniques.

For vehicle access rising kerbs and bollards can be employed, although it must be understood that these give no protection to the restricting of pedestrian traffic. If low security access is needed to control pedestrians with vehicles there is a range of barriers available and these are all scrutinized in this chapter.

In real terms there exists a huge range of equipment that can be used as physical hardware to satisfy the client and the application.

Tokens and credentials

The token is effectively the component that is used to identify the user to the access system. In electronic access control the term token is also referred to as a credential. In broad terms the token or credential is the document that verifies a person's identity whilst the person who presents the credential to the system is classed as being authentic. These credentials refer to cards, tokens, proximity tags and physical behaviour patterns or biometrics. A token may also be regarded as a sequence of numbers or a PIN (personal identification number) typed into a keypad. As with any industry there are different terms used but whereas cards and tokens are presented to media readers for authentication, the biometric elements are classed as being verified. When any of these is validated it is at that stage that access is granted.

Controllers and readers

The controller or reader is that part of the hardware that determines whether the credential is valid. The degree of complexity depends upon the token or credential and ranges from a basic tumbler mechanical lock to a computer system at a separate location. The controller is generally installed close to the access point in a distributed intelligent system or connected to a computer at a separate location in a centralized system. Some systems are a combination of the two approaches. The reader and controller may indeed be one integrated unit, as in the more basic lower security system, or they may be installed at separate locations with the reader being at the point of entry and to which the token or credential is presented. If multiple readers are used they may all be connected to one controller that is located in a more secure area.

The access control system can now be said to consist of four essential parts:

- A secure perimeter: this covers the walls and fences plus the physical restraints and boundaries.
- Access point hardware: the lockable and movable gates, barriers and doors plus other automated restraints.
- Tokens and credentials: the components used to identify the person seeking entry.
- Controllers and readers: the system part which verifies the request for entry. It may also be computer based.

In addition to the above certain systems will also have a human security presence in the form of security guards or perhaps a receptionist who can interrogate an entrant from a remote point and then grant entry. These may also be used to help users who attend on a casual basis and are not familiar with the system use.

It can also be confirmed that a secure perimeter covering the physical restraints can be supplemented with other security system concepts such as intruder alarms, CCTV and lighting. These can give outputs to show detection of a body attempting to gain access through an unauthorized route. Indeed the secure perimeter should be the first consideration of the access system. For many large premises and sites the access control system may well start at the boundaries of a site that are well in advance of the site buildings. This is intended to deter intruders at a point remote from the site buildings and which can be at some considerable distance from the premises. Reasons for access systems include the security of authorized persons and extend to the prevention of theft, vandalism, espionage or sabotage and terrorism. Access systems lend themselves not only to restricting access but they can also restrict the exit from a premises by which staff may attempt to remove stolen goods.

The access point hardware is very much governed by the level of security required and whether it is intended to automate existing restraints or to install new purpose designed equipment. In practice any existing lockable restraints can be automated, although in certain cases it may be more cost effective to introduce specific gates, barriers or such that have a particular history of being controlled by an electronic access control system.

The time honoured token or credential is actually the mechanical key which operates a lock but these are disadvantaged in that if a key is lost or stolen then for complete security the lock should be replaced. With electronic access systems this problem is overcome by encoding new data or removing old tokens and their identity. New tokens are also easily introduced.

1.3 Barrier types

As already stated site security can effectively be a combination of physical measures to include walls and fences plus any electronic security such as intruder, CCTV and lighting. These measures are needed before determining the requirements of the access control system itself. The perimeter itself must be secure in relation to the level of access security needed.

In practice almost any barrier may be automated so that existing gates that were originally purpose made can be adopted into the access system. Although all vehicle barriers do delay unauthorized access they do not necessarily prevent it and it must be understood that for a barrier to be highly secure it must have continuous surveillance. Lightweight gates are indeed often used if ice can impose a hazard or if there is concern that a vehicle could be hit from a following vehicle causing it to collide with a barrier.

There are certain barriers in more general use and those that are historically the most popular are described as follows.

Swing gates

These tend to be used to control vehicle access and require a large flat surface on the side of the gate through which it must swing; because of this they are not practical for some applications. There is a large amount of movement needed to operate a swing gate and the opening/closing times can be excessive with a general figure of 15 seconds applying for a motorized gate of average span. Indeed many motorists could feel intimidated if the gate movement was too rapid when the gate was opening towards them. A further consideration is that space must be

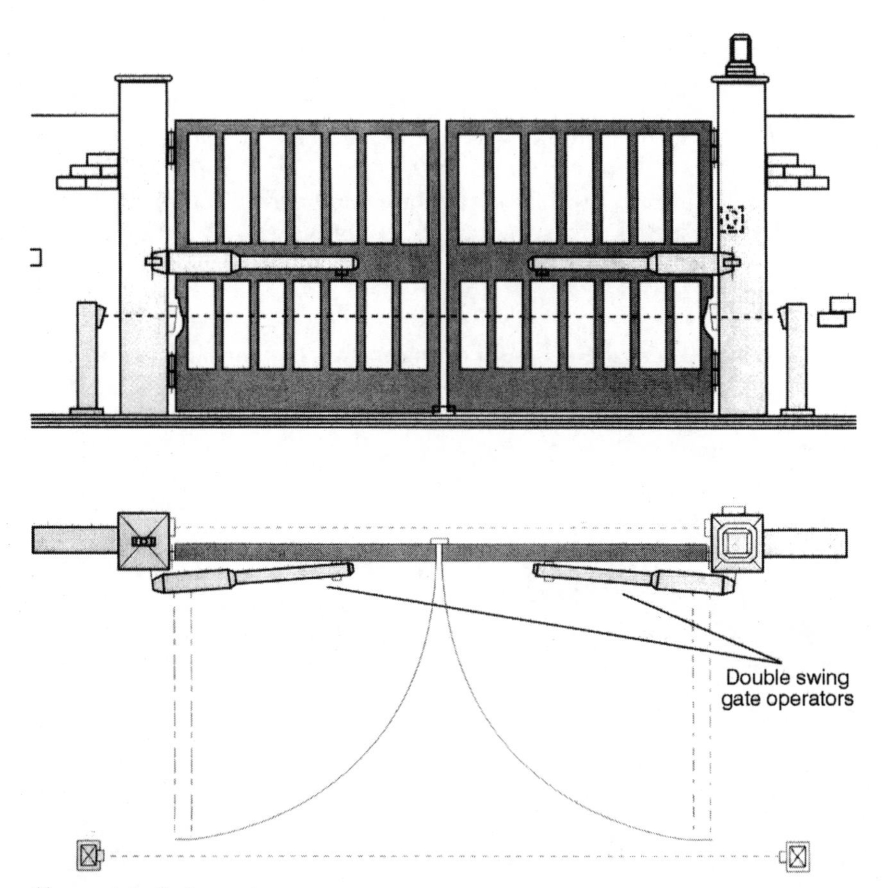

Figure 1.1 *Swing gates*

allowed for the opening arc of the gate and waiting vehicles, and the reader must be correctly sited at the position at which the vehicle is held for verification.

Although it is possible to use gates which fold as they open and save space the most used option is to install double smaller gates which save a great deal of space in the opening arc. The only disadvantage lies in the greater cost of the double gate, the installation of two units rather than one and the extra cost of the electric drive or hydraulic unit that powers them. Swing gates also have low impact resistance but remain popular for low to mid risk security applications (Figure 1.1).

Sliding gates

Also used in vehicular access control systems sliding gates are found in one of two types and are used when space is at a premium behind the gate. They tend to be seen sliding parallel to security fencing and walls.

Tracked This form of sliding gate features runners that are guided along a steel track that is formed in concrete in the surface of the ground. It is extremely strong as its weight is supported in the track and the only main consideration is that of ensuring the track stays clear of obstructions and is periodically cleaned. It has a much greater impact resistance to that of the swing gate.

Cantilevered These are self-supporting in that they are held at one end only by a strong supporting framework. They are disadvantaged in their reduced resistance to impact, although maintenance is reduced since there is no track to keep clear (Figure 1.2).

Rising arm barriers

Although swing and sliding gates can be used to control the access of both vehicles and pedestrians, the rising arm barrier can be employed if pedestrian traffic is unimportant since it provides a quicker means of operation and at a low cost. These are sometimes also called pivot gates or barriers.

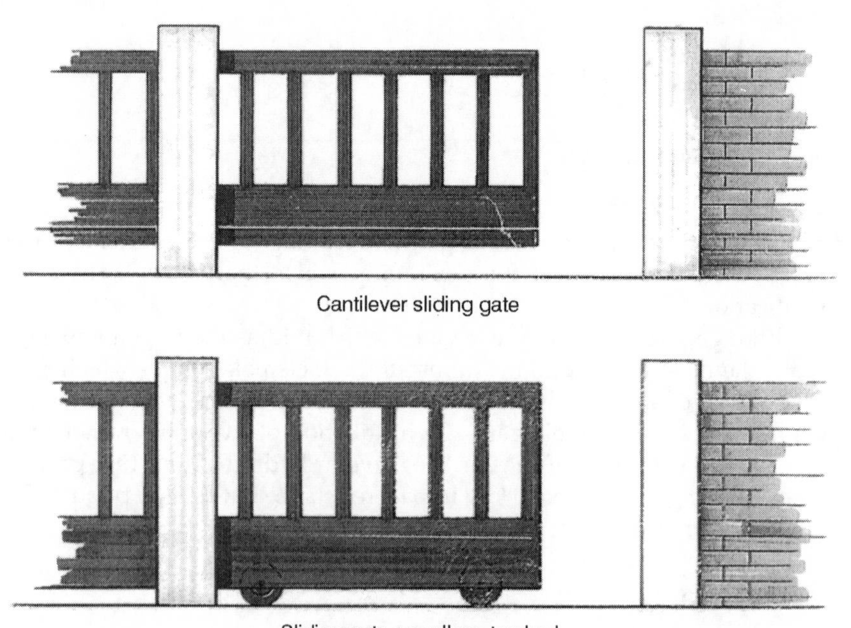

Figure 1.2 *Sliding gates*

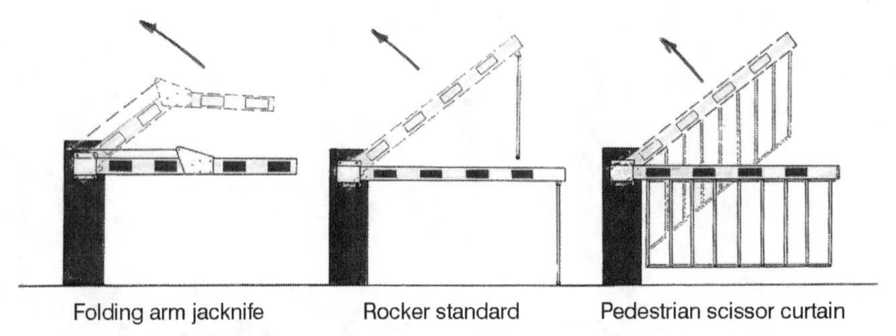

Figure 1.3 *Rising arm barriers*

Standard and industrial strength versions are available although neither provides high impact resistance and indeed the arms may be designed to break on impact to reduce vehicle damage. The industrial version has a longer span and is capable of withstanding a greater level of abuse and a certain degree of impact.

In underground car parks 'jack knife' arms can be employed as these need less headroom to operate. In the event that pedestrian access is required metallic curtains can be added to the barrier. These collapse as the barrier rises so that there is little obstruction to high sided vehicles. In the down position they give good resistance to pedestrian penetration.

Rising kerbs and bollards

These elevate from the road surface to restrict vehicle access only and are often also called pop-up barriers. They have high levels of abuse to impact and are rapid in operation. These devices may also be increased in number to work in tandem or to form a wall of protection.

They can be used in landscaped areas and business parks but must be used in conjunction with 'stop' and 'go' lights due to their low visibility. Switches are used to confirm that the unit is retracted and drainage must be installed to take away water from the underground mounting positions (Figure 1.4).

Turnstiles and paddle barriers

In Chapter 2 we discuss door types as these are the most used and traditional method of restricting pedestrian access. However, for external perimeter protection we must first look at tail gating or piggybacking as this is a barrier technique.

Although door closers and staff training can help to counter tail gating, in order for it to be truly effective mechanical protection may also be

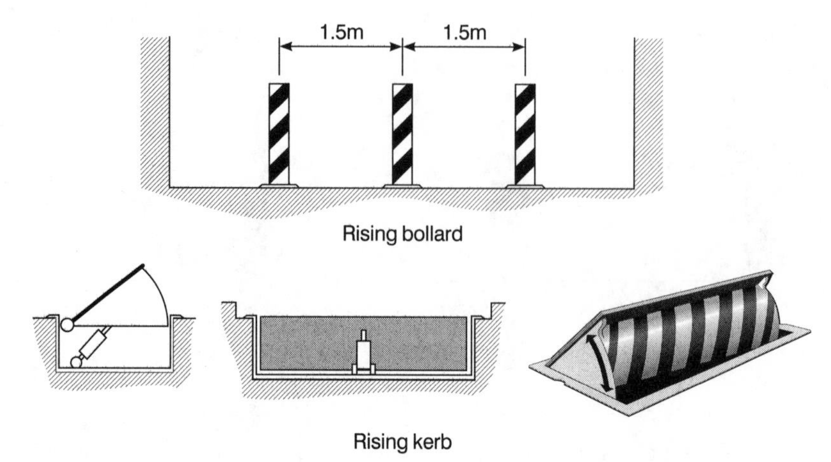

Figure 1.4 *Rising kerbs and bollards*

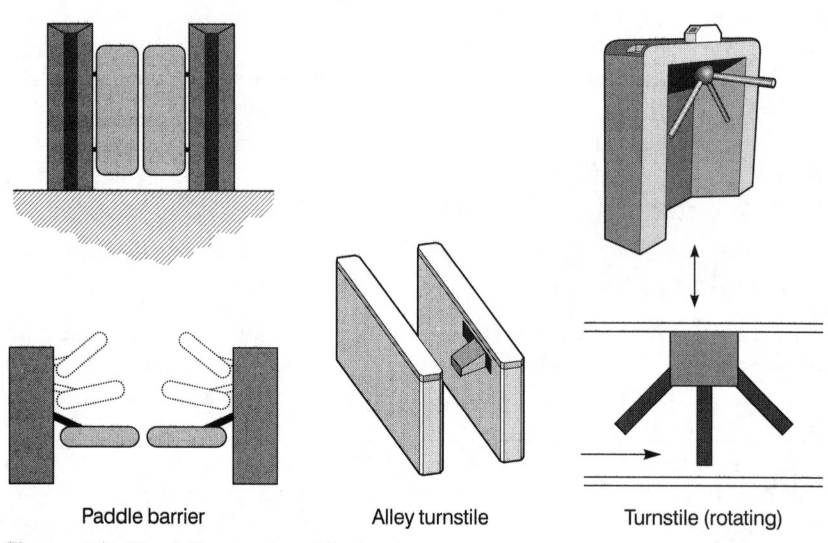

Figure 1.5 *Turnstiles and paddle barriers*

needed. The normal method of achieving this is through turnstiles and paddle barriers as shown in Figure 1.5.

This form of barrier decreases casual piggybacking by forcing people to enter one by one and is further enhanced by the use of alleys or narrow lanes. These lanes are only able to hold one person at any given time. For high security risk entrance ways, secure vestibules and turnstiles may be used as these have glass portals that hold persons in a sensing area enabling them to be visually verified before access is granted.

The actual barrier type will be governed by such factors as aesthetics and whether a security guard is also present. In the event that barriers can be vaulted a movement detector can be used to monitor the area above the barrier. This sensor is only enabled if the barrier remains closed. For external operation where aesthetics are unimportant full height barriers can be used.

Before selecting turnstiles and paddle barriers further considerations must be given to persons in wheelchairs or carrying parcels and cases as separate access points are then needed.

Air gates

This is a low security pedestrian or vehicle entry system without any physical barrier that notifies security personnel at a reception point if access is made through a beam detector providing that the entrant does not have a token capable of being verified by a radio frequency (RF) receiver unit.

The system comprises a beam detection unit which senses both pedestrians and vehicles, and aerial loops buried in the ground of the access point that sense RF transmissions and authorized tokens (Figure 1.6). The operation is as follows:

1 On crossing the buried aerial the pedestrian or vehicle is interrogated to ensure a valid token is held.
2 Providing a valid code is read the beam detection unit is disabled for a prescribed period enabling entry to be made.
3 If a valid code is not received the beam unit gives an output to alert security personnel.

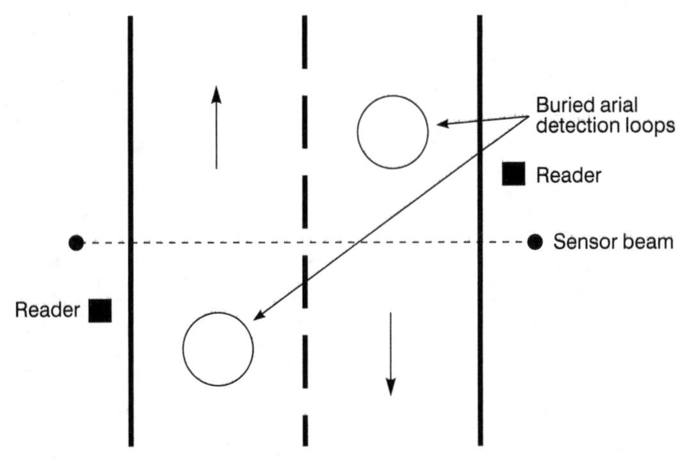

Figure 1.6 *Air gates*

In addition to the air gate method there are also low security applications such as in the residential care environment in which automatic gates that are normally held closed will open automatically to every approaching vehicle or person. A movement detector is used to signal to a reception area of the approach and this may perhaps activate an observation camera. This is not true access control as no verification is made to assess the authenticity of the approaching person or vehicle. However, it does have merit in stopping residents from vacating the perimeter of a care home whilst not intimidating visitors.

We are now able to summarize the barrier types as at Table 1.1.

Table 1.1 *Barrier types. Capability data*

Barrier type	*Stops vehicles Y/N*	*Stops people Y/N*	*Opening time (av. secs)*	*Span typical (min. – max.)*	*Impact resistance High/Med./Low*
Standard strength swing gate	Y	Y	10–25+	2–10 m	low
Industrial strength swing gate	Y	Y	10–25+	2–10 m	med. to high
Tracked sliding gate	Y	Y	10–25+	2–15 m	high
Cantilevered sliding gate	Y	Y	10–25+	2–8 m+	med. to high
Standard strength rising arm barrier	Y	N	2–5	3–5 m	low
Industrial strength rising arm barrier	Y	N	6–8	3–10 m+	med.
Rising arm barrier with pedestrian curtain	Y	Y	6–8	3–10 m+	med.
Rising kerb/bollard	Y	N	2–3	2–5 m	med. to high
Heavy duty rising kerb/bollard	Y	N	4–5	3–6 m	high
Turnstiles/paddle barriers	N	Y	1–2	1 m	N/A
Air gate			Refer to text		

1.4 Barrier configurations

Having gained some knowledge of barrier types it will now be understood that in addition to technical work there must also be a measure of civil engineering work undertaken, so certain skills are necessary. This applies across the range of barriers from automated gates, rising barriers, bollards, rising kerbs or any other physical devices. On the basis that the equipment may also need to be connected to the electrical mains supply, unless low voltage drives are used, the installer must have an understanding of current legislation and safety guidelines, but with some gates a knowledge of certain other techniques and how they are applied will also help. For instance some understanding of welding and the applying of concrete and construction bases for the larger installation

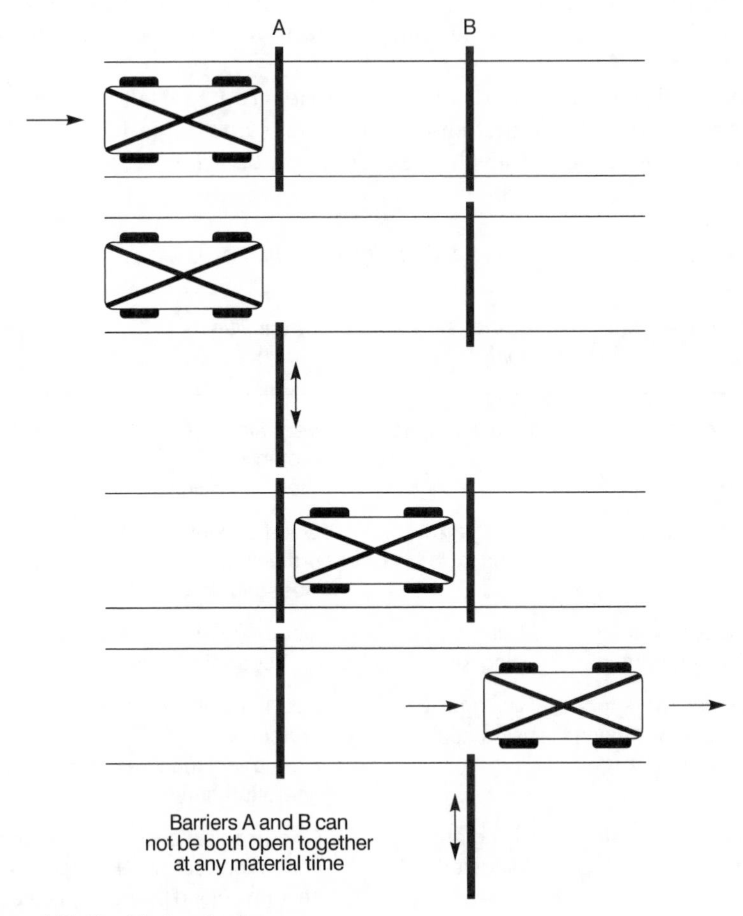

Figure 1.7 *Double barrier trap*

would be an advantage although more specialist help can always be sought for this part of the work.

Once an understanding is gained of the barrier types that are in general use it is then possible to identify the way in which all these barriers can be configured.

The double barrier trap

This stops tail gating or piggybacking and is used when vehicles are to be held and processed. It also enables vehicles to be removed from the public road and onto the perimeter of the protected area. It is effectively two barriers within one lane and opening in such a sequence that both cannot be open at the same time. It is highly secure as Figure 1.7 illustrates. The double barrier trap is often called an air lock technique in North America and in countries that follow United States and Canadian practices. It is not to be confused with air gate methods as these have no physical restraints.

In practice any similar or different barriers can be used in conjunction with each other at the first and second points and Table 1.2 shows as an example how different levels of security can be achieved.

Table 1.2 *Double barrier traps. Security levels and capabilities*

Barrier first point	*Barrier second point*	*Capabilities*
Industrial gate	Industrial gate	Very high security and impact protection. Slow operation
Industrial gate	Rising arm or rising kerb	Good security and impact protection. Reasonable speed of operation
Rising arm with pedestrian curtain	Rising arm or rising kerb	Good security, impact protection and operating times
Industrial strength rising arm or heavy duty rising kerb	Rising arm or rising kerb	Good vehicle security and impact protection. Little pedestrian security. Good operating times
Rising arm or rising kerb	Rising arm or rising kerb	Good vehicle security. Reasonable impact protection. No pedestrian security. Good operating times

The next three approaches – 'common lane', 'single entry and exit lanes' and 'dual entry and exit lanes' – allow a more rapid traffic flow to be achieved. With these approaches a gatehouse is generally used so that delivery and visitor access can be incorporated into the system by manual vetting whilst staff may be processed more quickly by presenting tokens to an automatic reader. The gatehouse tends to be sited between the lanes so that high vehicles cannot obscure the view of security personnel from being able to view all of the lanes at any given time.

It must be understood that when the system is used with a reader only to clear staff vehicles, no authorization is a made of the staff in the vehicle. The system is unable to verify the number of people being transported in the vehicle or exactly who these may be. The authorization is hence given to the vehicle via the token only.

Common lane

A double barrier trap is highly secure but it is unable to operate where high traffic flow is experienced and for this application a common lane is used but operating as a two-lane roadway governed by a single barrier. It is inexpensive but low security since unauthorized traffic or personnel can also enter or exit when the barrier is opened to an authorized user, as in Figure 1.8.

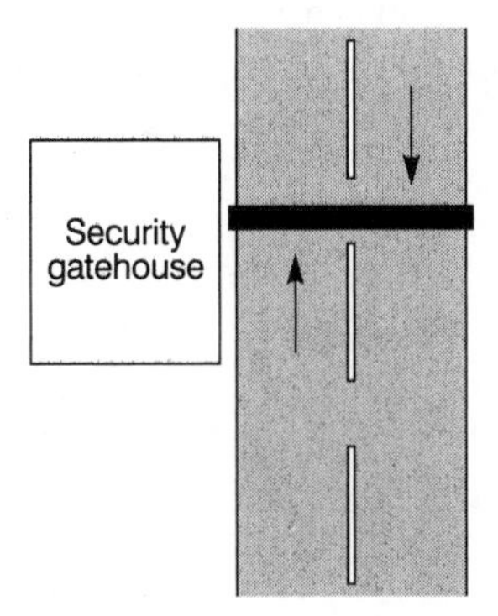

Figure 1.8 *Common lane*

Single entry and exit lanes

In this system the gatehouse tends to be found sited between the entry and exit lanes with individual barriers controlling each lane. This method is in practice widely used as it gives a good level of security and is easily managed (Figure 1.9). Since visitors and delivery vehicles must be screened and hence the traffic flow is slowed as only single lanes are

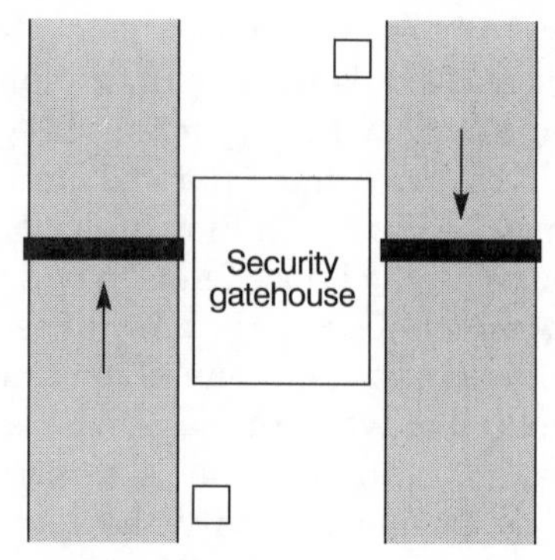

Figure 1.9 *Single entry and exit lanes*

used, some consideration must be given to this timing to occur during periods of low staff traffic. If this can present a problem dual entry and exit lanes are advisable.

Dual entry and exit lanes

This is similar to the previous technique but has separate lanes for both entry and exit of staff, delivery drivers and visitors. It is a more expensive option but is particularly useful in dealing with high traffic flows. It may be possible to run the system without a need for a dual exit to enable all users to exit by one lane or staff to exit by their entry lane. The width of

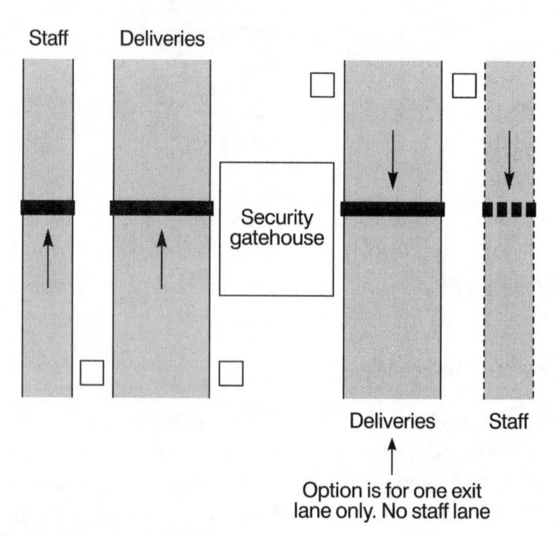

Figure 1.10 *Dual entry and exit lanes*

the lanes is governed by the vehicles attending the site. The delivery lane may need to be considerably wider to permit heavier vehicles, unlike the staff lane which is intended for cars only (Figure 1.10).

1.5 Safety considerations and emergency exiting

Safety remains paramount in access control and when dealing with barriers it is clear that there are certain safety measures that must be undertaken. This extends from the safe closing of the barrier to that of allowing safe emergency exiting to be achieved.

Safety rules always require that people must be able to leave an area safely even though precautions have been made to make it difficult for entry to be granted. All barriers used in public buildings are regulated by both legislation and building regulations as they influence emergency exiting.

Safe opening and closing

As the barrier opens and closes it must not cause any damage to people and vehicles passing through it. Although barriers can be so designed as to break on impact during their opening and closing cycle if they are to hit an obstruction, this ideal may not always be suitable since some barriers must have high degrees of impact resistance. Therefore the normal method of incorporating safety is to fit sensors at the barrier edges. These may be simple switches that automatically stop the barrier when an obstruction is sensed. An alternative to this is a road loop that senses objects in the path of the moving barrier and stops its motion until the obstruction is removed.

In all cases these safety features are best incorporated directly under the control of the local barrier in question rather than at a central point which could control a number of barriers and functions. This is to ensure that it does not compromise the safety measures of the local barrier.

Emergency exiting

This may be required if there is a power failure or an emergency that demands the clearing of a site. It is possible that in the event of an emergency all barriers can be given an electrical input so that they all physically open. All barriers can be designed so that in the result of power failure they are held open, kerbs and bollards are retracted and turnstiles become free to rotate. There are also options to provide manual overrides so that all barriers can be manually opened to allow exiting to be safely carried out. Indeed in the event of power failure manual overrides are needed in the design so that normal entry of personnel to a site can be made.

The barriers used across the range of access control systems will vary enormously in terms of strength, security and means of operation. Ancillary devices will include internal and external photoelectric cells to check that the area in which the gate operates is clear of any obstruction. For additional security and safety we will also see flashing lights, loop detectors and alarm interfaces. There may also be emergency power supplies and acoustic operating signal equipment.

Electrical characteristics will also vary. As an example, swing gate operators can be attached to almost any existing gate for single or double leaf operation. The lighter application operator can be powered by a 12–24 VDC system or a single phase 220/240 VAC supply with a motor capacity in the order of 250 W with thermal safety intervention.

Sliding gate operators may require a mains single phase supply with a rating up to some 350 W or operate from the three-phase 415 V supply with a rating in the order of 300 W.

Heavy duty industrial and traffic barriers or bollards may use single or three-phase motors with the motor protected from overheating by a single or double thermal probe to interrupt the supply. Ultimately the drive may be hydraulic or worm gear with electrical and mechanical limiters. In the event of a power cut the geared motor can be disengaged to enable manual override operation.

The controller as supplied with many barriers and operators is an integrated and purpose designed component. However, it remains to say that any barrier can be automatically driven by any controller if interfaced with the correctly rated control gear and contactors or relays.

1.6 Discussion points

Barriers define a clear boundary and delay vehicle and human traffic whilst directing attention to the designated entrance points. They vary in the levels of security they afford but to be highly secure they need constant surveillance. Barriers are also governed by many safety features, including emergency exiting and monitoring as appropriate, to ensure that they cannot impact obstructions within their movement path.

Most existing mechanical barriers can be automated but doing a survey a new perspective is introduced when selecting the most efficient operator in preference to specifying a purpose produced and time-honoured barrier that comes complete with all of the specific control gear.

In all cases there is a need to provide cabling to the barrier points. These could operate from a multitude of sources ranging from 12 VDC through to single- and three-phase mains supplies.

There is also a degree of construction work associated with every installation and this may well be diverse in relation to that in which the traditional electronic security installer is familiar.

All of these subjects are to be considered and discussed at any survey before the most efficient and cost-effective solution will be found for any particular application.

2 Doors

In Chapter 1 we started our investigations into perimeter protection and then considered barriers and the different types and configurations that are in current use. Similarly, doors can also be classed as perimeter protection devices. Indeed they form an equally large subject as that of barriers because their construction also governs the locking mechanism that can be used.

In this chapter we assess the different types of doors plus their construction and supporting framework. We then consider the locks that are used to electrically control them plus the supporting hardware.

Monitoring of the door for its orientation is a normal requirement as is the status of the locking condition. These devices may also be subject to vandalism and attack and so must be considered as well. In addition there is a need to recognize the requirements for emergency exiting and escape routes.

There are a number of techniques involved with all of the locking functions as well as different wiring methods. This chapter will give the required insight into all of these considerations.

2.1 Door types

When controlled by an electronic access control system all doors will be found to have certain common features:

1 An electronic means of locking and unlocking a door. This may also be performed by an electromagnet. The way in which a door opens also influences the choice of the locking mechanism.
2 A mechanism to close the door to ensure that it is normally held closed and locked.
3 Sensing devices should also be used to determine the closed or open state of the door for security purposes.

In the first instance it is prudent to determine and define how we can actually class a door type.

The hanging of the door is always determined by the position of the hinge knuckles when the door is closed. Remember to stand with the door opening towards you (Figure 2.1).

- Standard doors – push to open and enter.
- Reverse doors – pull to open and enter.
- Swing doors – capable of opening in both directions.
- Double leaf doors – may open in one direction or operate as swing doors. Effectively two doors in one frame.

Construction – door and frame

This very much depends on the level of security and also on the type of lock preferred, as the locking mechanisms differ between those used on wood, metal and glass doors. Equally aesthetics plays a part if cables must be run to the lock through the door and if it is difficult to take them through or hide them within the door itself. Metal frames and doors can in some situations also cause interference with RF readers.

Environmental conditions must be considered. Solid wooden doors expand in damp conditions and steel doors will show a measure of expansion and contraction. This also applies to the frame that supports the door.

The greater strength offered by the door and frame enables it to better withstand vandalism and attack. In cases when this is a distinct possibility it is advisable to employ solid doors of reverse operation of single leaf only because they are surrounded more strongly by the frame and it is more difficult to attack by pull forces than by push forces. Double doors are not recommended and heavy hinges made from brass or steel should be used with the frame bolted to the surrounding brickwork. Wooden doors should be no thinner than 44 mm and hollow doors are not to be used for external applications.

The door construction also governs the selection of the lock. Electromagnetic locks or electric strikes are needed on glass doors due to the difficulty of cabling an electric lock if it is to be sited on the moving door in preference to the fixed frame. Solid wood or steel doors also present the

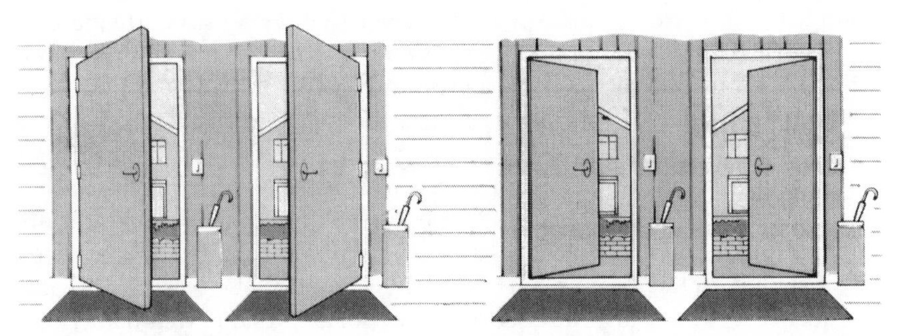

Figure 2.1 *Door types*

same problems, whereas interior doors of hollow construction are more easily accommodated as it is relatively simple to run cables in the door cavity. To summarize, some electric locks are fitted on the frame whilst in other installations the locks are installed on the door. It is always more easy to take cables to the frame only without having to loop them to the opening side of the door so a careful lock selection must be made with respect to the exact door construction. The electrical signals can be taken to the opening side of the door by flexible door loops but these are avoided if possible.

Fire doors must be considered separately as the door operation could be degraded if its method of locking is modified; this can also affect its fire resistance. In practice the rules surrounding fire exits and doors sometimes conflict with access control systems and in general exits for this purpose must be:

- Easily visible.
- Simple to operate and by one action.
- Designed with minimum hardware using one opening device with a smooth surface.
- Not be locked from the inside.
- Have an automatic door closer.
- Be constructed of fire rated material.

In addition to these requirements consideration must be given to people in wheelchairs, the visually impaired and to the infirm. The safety of these people is not to be compromised.

Door closers

Any door that does not self-close is to be fitted with a door closer to perform this operation. It is not acceptable to rely on either staff or visitors to carry out this task when entry or exit procedures have been followed. Door closers are spring activated with a tension adequate to fully close doors after use yet not so strong as to make opening the door difficult. Both concealed and surface mount versions can be found.

Concealed door closers have the arm or mechanism attached to the spring which guides the door shut embedded into the frame and door itself. These are more often found purpose made for the door type and installed at the point of manufacture. Surface mount door closers are more popular and are often added to an installation at a later stage. They are installed on the interior side of the door for aesthetic reasons and also because it reduces any tampering or exposure to the weather. These devices are easily installed and the spring force can be adjusted to suit the application. The opening force is generally adjusted to cater for the person of light physique and for those carrying parcels or in wheelchairs. The closing speed should be adjusted to ensure that the door latches but does not clash and although fast speeds deter tail gating this may need to be sacrificed.

Door closers will be found to have an adjustment screw that cushions the door at the point of closing and latching.

2.2 Locks

Traditionally locks have required a mechanical force to operate the mechanism; however, with electronic access control systems the mechanism is operated by an electrical technique. Nevertheless most electrical locks are based on a similar and time honoured mechanical lock form so that they are generally interchangeable; however, cabling needs to be provided when installing an electrical lock. Many electrical locks will also have a mechanical override. The two most widely adopted locks are the mortice lock and the rimlock.

Mechanical locks

Before considering the functions of an electrical lock it is important to overview the standard mechanical lock. There are three varieties: latch, dead bolt or a combination of the two (Figure 2.2).

For locksmithing purposes the lock is a device that incorporates a bolt, cam, shackle or switch to secure an object such as a door to a closed, locked, on or off position and that provides a restricted means of releasing the object from that position.

Once the handle is turned in the follower the latch bolt retracts against the spring force of the latch springs. When the handle releases the pressure, the latch returns to its original position effectively locking the door. The lock shown also has a dead bolt enabling a key to lock the dead bolt in position.

The dead bolt can only be retracted by a key and not by turning a handle. Figure 2.2 clearly illustrates the popular lock with both a latch bolt and dead bolt with the latter providing additional security to the handle operated latch. This technique as described is adopted by the mortice lock that is widely used in the building industry.

- Mortice lock. These are cut into either a wooden door or the frame or are mounted within a cavity of an aluminium, PVC or steel door. They possess good levels of strength being held within the supporting material and as they are not visible the aesthetics of the installation is enhanced (Figure 2.3).

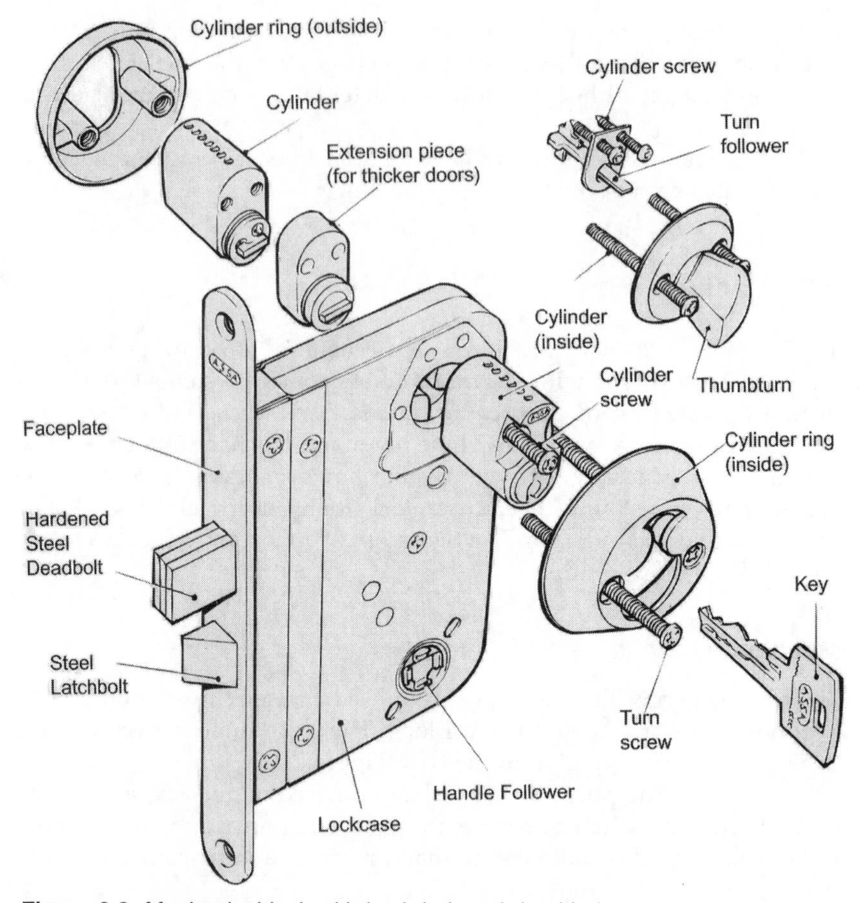

Figure 2.2 *Mechanical lock with latch bolt and dead bolt*

- Rim lock. These are different to the mortice lock and use an alternative technique, being surface mounted on the inside of the door with the strike plate or lock fixed to the door frame. They are low security as they are not cut into the structure, but nevertheless they are easy to install and understand so remain widely adopted (Figure 2.3).

Electrical locks

These operate by the switching of an electrical current through the lock as provided by the access system controller. The electrical lock combines the strength and quality of the mechanical lock but with the flexibility and convenience of electrical control. They offer superior operational security than the electric strike or release to which we turn our attention later.

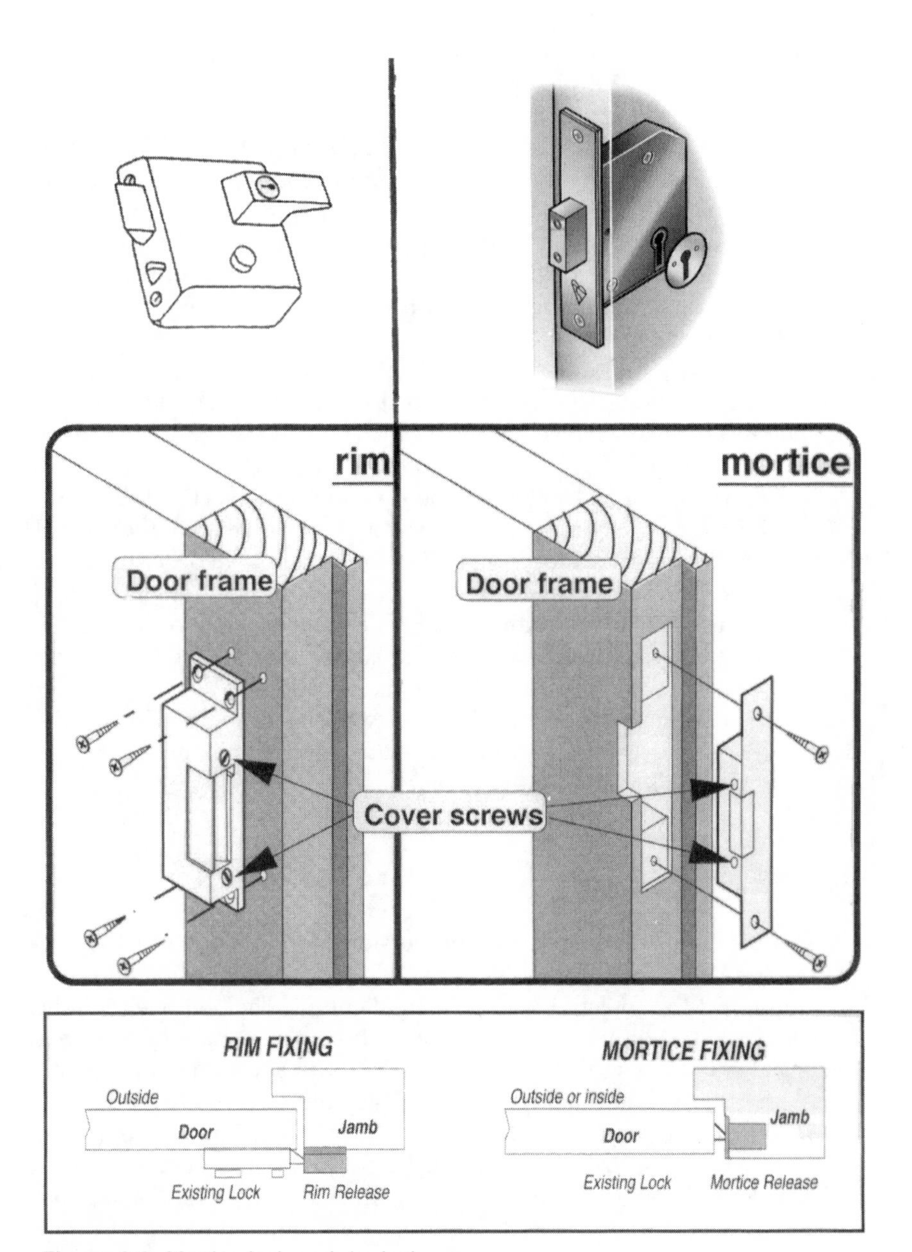

Figure 2.3 *Mortice lock and rim lock*

In practice the electrical lock is one of two versions:

- The electric lock controls the function of the door handle.
- The electric lock controls the lock bolt directly.

The electric lock also tends to have a mechanical function on the interior side of the door for emergency operation, in the event of a fire or to allow the door to be opened if a power failure occurs and there is no standby electrical source. Equally it is often convenient to use a conventional handle or key on the inside of the door for normal exit procedures rather than using a 'push to exit' egress button that provides an electrical pulse to release the lock.

Electric locks may be either locked or unlocked when the power is removed. They may use a solenoid to disengage the handle from the follower mechanism which controls the latch bolt and so are similar to the lock shown in Figure 2.2. When locked the handle can be turned but the latch bolt will not withdraw since there is no mechanical link between them. However, when unlocked the latch bolt will be in the locked position until mechanical force is applied to the handle in order to retract the bolt. The bolt will then relock under pressure from an integral spring as the force is taken away from the handle. However, the door must close with sufficient force to push against the latch spring to retract the bolt when it contacts with the door frame striker plate.

Single or double cylinders may be used with electric locks to enable the door to be opened from the inside and outside with hook bolts adopted for narrow stile doors or for doors that slide.

Electric solenoid bolts

These are a heavy duty compact unit version of the electric lock and are mainly used in metal frames. They have lock status monitoring and tend also to protect the solenoid with a thermal cutout. The bolt itself will be bevelled on its holding face to prevent binding in the strike under abnormal side loading of the door, caused by door warpage or persons attempting to lean on the door and trap the bolt.

An auxiliary spring-loaded plunger actuates a switch when depressed by contact with the strike that allows the bolt to be projected.

The solenoid bolt can be used in 'panic' locking applications that take place in the event of a potential threat, e.g. petrol station forecourts, cash doors and jewellers' shops. These provide an instant lock facility from a manually operated switch.

Key cylinders or thumbturns can be found as optional extras on the inside surface to enable a manual override to be achieved.

Electric strikes/releases

The electric strike is still the most widely used method of dealing with the control of an existing lock. They are available in a range of ratings and styles and can be specified for continuous duty in which they are energized for long periods of time or they can be specified with different size face plates depending upon the frame into which they are intended to be mounted. If flush mounted they tend to be called mortice lock releases and if mounted on the outside of the frame they are called

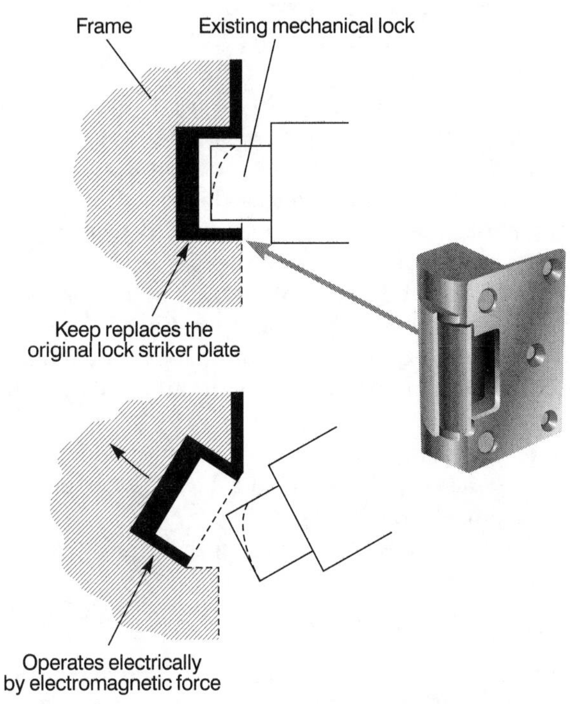

Figure 2.4(a) *Operation of electric release/strike and door keep*

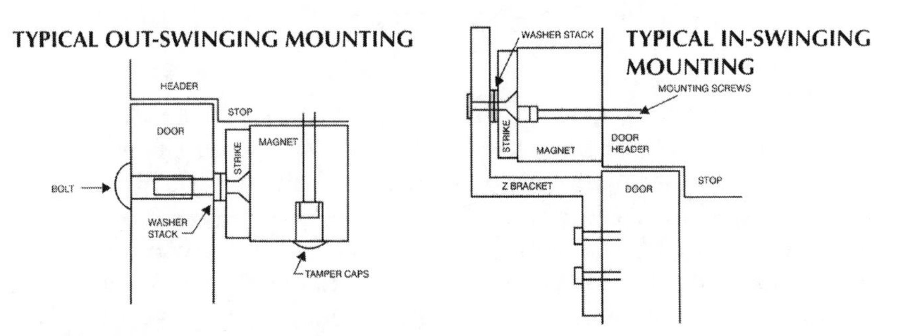

Figure 2.4(b) *Electromagnetic lock*

surface lock releases. They may be fail locked or fail unlocked as described later in Section 2.3.

In practice the terms electric strike and electric release refer to the same device with both the electric strike and release engineered to fit into a door frame.

The electric strike or electric release both release the keep on the door frame striker plate allowing a fully extended lock to pass through the keep without the lock retracting. This is illustrated in Figure 2.4(a), which is in contrast to the operation of the electromagnetic lock shown in Figure 2.4(b) and discussed later. The electric strike is designed for fitting in the door frame and to replace the existing lock striker plate, although this may involve a certain amount of cutting of the frame. However, the original mechanical lock continues to be used if a compatible strike is selected. Therefore we can say that the electric strike operates by allowing the keep of the lock to swivel into an open position when it unlocks therefore enabling the lock to be released. It is achieved by the use of a small electromagnet applying the force needed to retract the bars that secure the striker plate of the keep in position.

Prior to installation the strike plate should suit the strike and door as special strike plates are available for narrow stile doors. These devices are supplied as the standard unit with most door entry and low security access control systems and remain in use throughout the industry in great volume.

On installation a clearance of 1 mm between the lock bolt and release when it is in the locked position should be allowed with invertor cases used for outward opening doors.

Delayed action locks can be used to trigger the release to allow the door to open with the release being held in that position until the door has been both opened and closed. The release then reverts to its original state.

For door monitoring purposes monitor switches including both NO and NC contacts can be used. When the release is triggered and the door opened, the contact operates making it a suitable device for connection to an alarm device or specific telephone input.

Motor locks

Motor locks are a different form of lock as they use a small motor to retract the lock bolt against the pressure of the bolt throwers making them particularly strong. They tend to feature an electronic timer to quantify the period in which the bolt is to stay retracted. A major consideration is that of incorporating the necessary electronics to ensure that the door is closed when the bolt is energized or it would actually stop the door closing. This is achieved by either an electronic sensor or a mechanical trigger which is pushed back into its housing in the lock when the door closes. This operation allows the lock to release and is somewhat similar to the mechanical latch lock type.

The motor lock technique is not unlike the electric dead bolt lock that may be driven by a solenoid to retract the bolt and is used on doors that swing in two directions and on double doors. The dead bolt may be fitted either into the jamb or frame of the door or on the door itself. Due to their high security they may be found in sets on doors with one device at the top and a further device at the bottom because to force the door they must be attacked simultaneously. This form of locking mechanism is designed primarily for unit concealment and tamper resistance as it is compact and hidden. A consideration is that the bolt does not tend to have a manual override and therefore it is not advocated for emergency routes.

Electromagnetic locks

These secure doors through magnetic force and always unlock when power is removed and lock as the power is applied. They are not used in high security applications unless back-up supplies are available but they form a good option on doors in which it is difficult to fit traditional electric locks, although they are more expensive.

They consist of an electromagnet and strike plate or locking armature. The electromagnet is installed on the door frame, usually at the top, and the armature which is attracted to the electromagnet is fitted to the door. They do not wear as they have no moving parts and have built-in door status with anti-tamper indication. Extremely reliable they are available with monitoring options that can sense any attempt to lower the holding power of the lock by inserting a foreign material between the armature and magnet. Optical alignment sensors ensure that the armature is correctly aligned with the magnet before the lock is energized. They can also sense any attempt to obstruct the magnets. These locks can be specified with holding forces up to 1500 lb for medium security risks but are less efficient on tall doors since the leverage forces that can be applied to them are so much greater. In practice there are two types:

- Direct hold. These are surface mounted on the secure side of the door.
- Shearlock. These are embedded within the door and frame, so are concealed. These also have mechanical interlocks activated by the electromagnetic field and this stops the sliding or shearing of the components. Observations would show that if the sliding action is removed from these locks the holding force may be effectively doubled. So they can be specified up to some 3000 lb.

Electromagnetic locks are often found on poorly fitted doors, on glass doors and on certain doors with small frames. In North America and in those areas where ANSI standards are followed there are three grades:

Grade One. 1500 lb. Medium security.
Grade Two. 1000 lb. Low security.
Grade Three. 500 lb. Normal force to hold a door closed.

The mounting of electromagnetic locks does differ but Figure 2.4(b) shows a typical technique and can be compared against the method applied to the traditional lock.

Spreader plates sometimes have to be used if the frame profile does not offer a secure fixing and for in-swinging doors a Z bracket can be applied although these tend to be part of the kit supplied with the lock.

2.3 Locking considerations and wiring methods

The efficiency of the access system perimeter protection afforded by the doors and the electric lock is governed by a number of factors including:

- Door construction.
- Door type.
- Level of protection against attack.
- Frequency of use.
- Electrical rating of locking system.
- Emergency operation and power down.
- Door and lock sensors.

Door construction

This also governs the lock that is to be fitted as it may be difficult to run cables within the door cavity or to form channels within the door if it is solid. For that reason it may only be possible to install the electric lock or to use an electric strike/release within the frame so that no wiring is taken to the door. If it is a necessity to install the electric lock in the moving door, it is achieved by door loops that carry the electrical signal from the fixed wiring taken to the door frame over to the moving door. These loops consist of two junction boxes connected together by a flexible loop with the junction boxes being fitted to the frame and the door. They are to be fitted on the secure side of the door. Many locks are only designed for use on certain wood or metal doors, and glass doors in particular present difficulties. In such cases and providing that the required security level can be achieved it may be in the best interests to adopt an electromagnet or electric strike to negate the problems of wiring within the door and installing the locking device there. The cabling and fitting of the lock would therefore all be achieved at the frame, which itself must be of a solid construction.

Door type

This equally restricts the type of lock that can be used. Single leaf standard, open-in doors are the most easy to accommodate and being the most popular it will be found that almost any lock fits them. With reverse, open-out doors it is vital that the strike is not fitted on the outside where it could be attacked. If double doors are employed and must be both locked and unlocked by the system then they are both to be stationary and in the closed position before locking. This can be guaranteed by time delays on the locks supplemented by specific double door locks or electromagnetic locks. If the doors are to open in turn because of a rebate this presents extra complications since they must close to an exact sequence and to control this function is difficult.

Level of protection against attack

With external doors the strength they exhibit is of major importance, so they must be of solid construction. A single leaf solid reverse, opening-out door will always provide the greatest resistance to attack when amalgamated with a compatible and secure lock.

Electromagnets are not to be used if an attack is probable since they unlock if the power fails or is deliberately removed. The requirement for internal doors is not quite so onerous and hollow frames may be acceptable in many cases.

Frequency of use

This is the volume of use to which the system is put. All locks can be rated in terms of number of operations per time period and life expectancy. It is then a question of determining the number of operations a system will be subject to in use. This must take into account all the usage for entry and exit and include the periods where users may leave and re-enter the premises throughout the day.

Electrical rating of locking system

In practice any electrical lock can be linked to any access control system but it may involve the use of additional power supplies and relays which are used as the interface between the controller and the lock. Every system will have a switched output intended to signal the electrical locking mechanism. In some instances the voltage supplied will be direct from the controller but in others it will be through an external auxiliary supply. Certain locks can be powered direct and are advocated for use with particular systems, whereas others will need an interface depending upon the rating of the lock and that of the controller's lock output.

Most access control systems have locking mechanisms operating at either 12 or 24 V which may be AC or DC. It is only possible to energize the locking mechanism when its supply voltage and the output voltage of the controller to the lock are similar otherwise the system cannot work or it could cause damage to the lock.

Many users prefer the audible signal generated by the AC lock to confirm its operating period, but buzzers can be located alongside DC locks to provide a sound signal. Rectifiers can be added to AC locks in cases when silent DC operation is wanted.

With all devices the current loading of the lock must not exceed the controller's lock output or it will attempt to pull excess current through the controller's electronics. The current loading of the lock is the amount of current required to hold the lock in an energized state. It is this current rating together with the operating voltage that must be satisfied by the controller or the auxiliary power supply in order to drive the lock efficiently.

It must be appreciated that the current drawn by many locks does momentarily exceed the specified current rating when initially energized. This excess current is called either the inrush or pull-in current and needs to be taken into account. The ratings of electrical locks vary enormously across the range and it is not possible to generalize but data on the rating is provided with every unit.

Electromagnets create high transient currents that may be ten times in excess of the normal rating or running current, but these locks tend to come complete with a unique stabilized power supply to satisfy both the inrush and normal running or rated current. The power supply is triggered by the access control system's controller and the current or load is then derived from the power supply. The power supply is hence an interface between the controller and the locking mechanism (Figure 2.5).

The other consideration is that of selecting a lock that satisfies emergency override conditions in the event that the system could be depowered:

- Fail locked. In the event that the power is removed the locking mechanism and door will be locked. Power is needed to unlock this type of lock and it is used for normal locking situations. It is sometimes called fail secure.

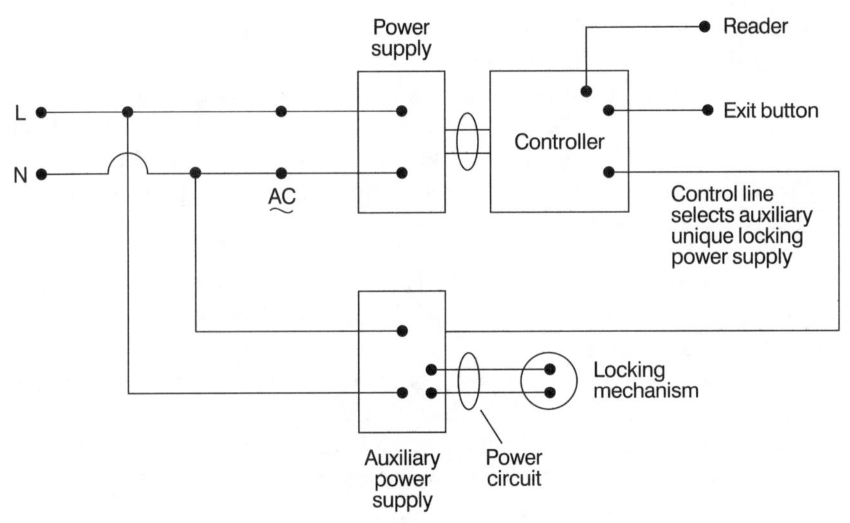

Figure 2.5 *Remote power supply*

- Fail unlocked. In the event that the power is removed the locking mechanism and door will be unlocked. These are often found on fire doors as they unlock automatically and are released in the event of the power being disconnected so are at times called fail safe.

AC locks are fail locked but DC locks can be found in either fail locked or fail unlocked modes.

Emergency operation and power down

This must be a consideration in the event that emergency evacuation of a building may be necessary or a disruption of power could occur or the power supply for the electronic access control system could become defective. It is also possible that the access door controller or signalling could fail or the electronics of the lock itself become faulty. Equally the cables could become damaged in a fire or under working circumstances. Mechanical overrides, such as a handle, panic bar or thumbturn on the secure side of the door are used to perform this manual function. It must be understood that with a lock that fails locked there are further considerations because the door would be rendered in a locked condition. The mechanical override will be found on the inside of the door with a key override on the outside. This external key override is needed in the event a power failure could occur when the building was vacant and would be needed to open the building from the outside. The key for this facility should be held in a secure location. It should also be of a special key design to prevent easy copying.

If fail unlocked systems are employed the electric lock is more complex as power must be applied to the lock at all material times other than when the door is being electrically unlocked. For this duty 'continuously rated' locks are needed which may mean additional power supplies or standby battery packs are required.

All fire exits and emergency doors must be easy to see, easy to open in one simple motion and not be locked from the inside. There must be minimal hardware used on these doors and this extends to the operation of any electrical system forming part of it.

It can therefore be seen that there is a need to respect not only the needs of security but also the conflicts that can exist with fire safety and normal access. Locks should not be viewed as a single device because they vary enormously in the options they feature and the ways in which they can affect the system as a whole.

Door and lock sensors

Door monitoring contacts assess the door's closed position and are used in conjunction with timers connected to a warning system to confirm that the door closes several seconds after each occasion that it is used. Timers are also used to stop doors being held open. The timer should be set to take into account the requirements for the older and infirm person or those carrying parcels. Sensors can raise an alarm if the door is opened without an authorized transaction being made such as the inserting of a card or PIN into the system. This prevents a door lock being deliberately disabled such as by trapping the lock bolt in a retracted position, although internal sensors can be used to ensure that when the door closes the lock bolt moves into the correct lock position.

Figure 2.6 illustrates the simplest form of monitoring. The single monitor switch enables the coil of the lock to energize only when the switch is closed whilst the indicators identify the monitor switch status.

Many electric strikes have dual monitoring accomplished by two sensor switches: one is activated by the latch bolt's penetration of the strike and the other by the solenoid plunger that blocks the strike's release. Figure 2.7 shows the arrangement.

It remains to say that the sensors form an important part of the door locking security. Indeed a vast array of options are available on all locks and systems and the functions will differ enormously between products and systems. In order to overview wiring techniques we can nevertheless study three case examples since these cover the essential methods.

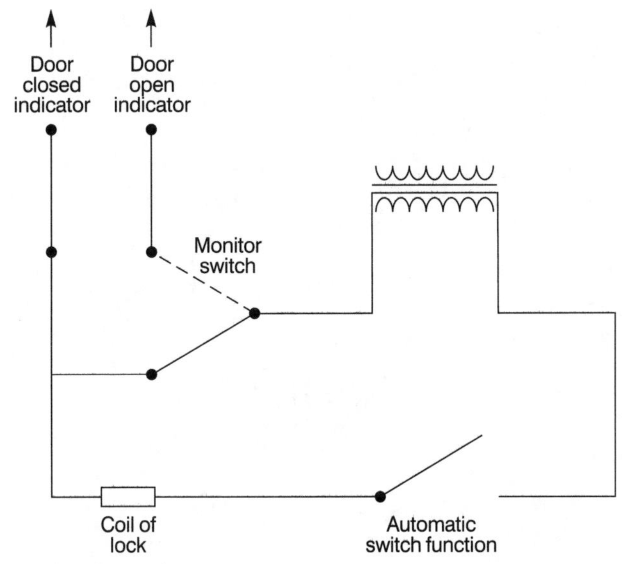

Figure 2.6 *Door monitoring*

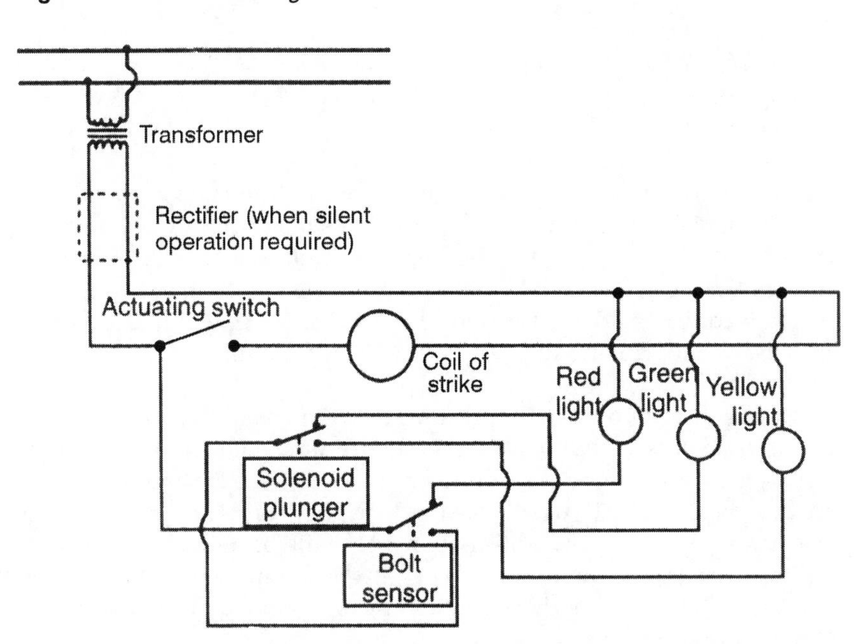

The green light indicates that latch bolt is in strike and solenoid plunge is blocking strike. Yellow light indicates latch bolt is in strike but strike is unblocked, allowing access. Red light signals that latch bolt is out of strike.

Figure 2.7 *Wiring monitor operation*

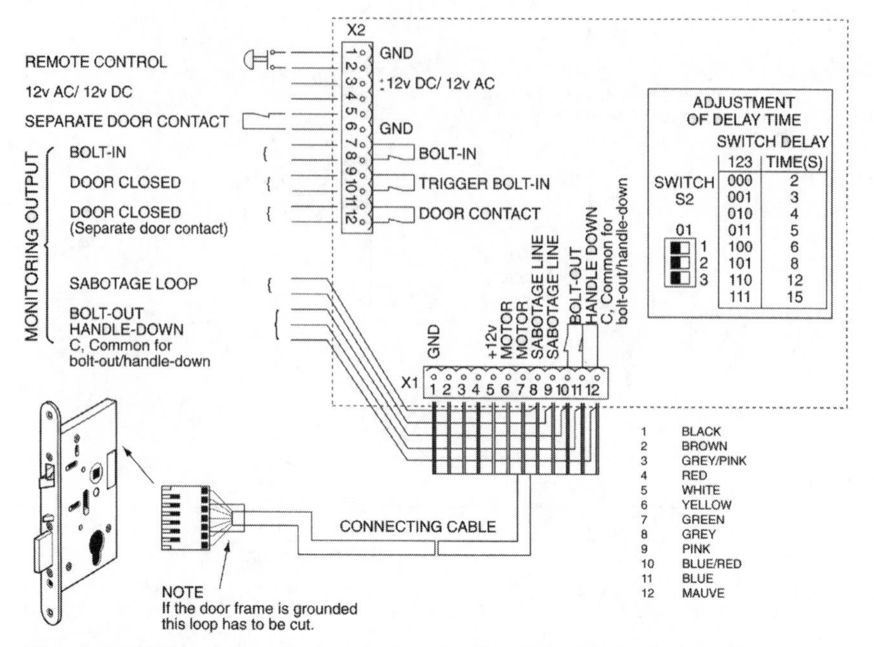

Figure 2.8 *Wiring motor lock with dead bolt and anti-friction bolt*

Example 1 A motor lock with a dead bolt and anti-friction bolt. With this version the control unit is separate and interfaced by a connecting cable to the lock assembly. The bolt is withdrawn by the motor and thrown by spring force. When the control switch is continuous, the bolt is retained in the lockcase and the lock stays open. If the control impulse is momentary, the bolt is deadlocked again after an adjustable time period delay.

If the bolt is thrown by depressing the trigger bolt when the door is open, the bolt is withdrawn by the motor immediately after the trigger bolt is released to prevent abuse of the lock. The bolt and the anti-friction bolt will be deadlocked when they are thrown into their outermost position. The anti-friction bolt will be unlocked when the bolt is withdrawn. In the event of power failure the bolt is thrown automatically but a mechanical override is provided. The wiring is as shown in Figure 2.8.

Example 2. In this example the available options are separate day/night functions in that during night locking the handle does not operate the bolt whilst in day it does operate. The switching from night to day is remotely controlled. This is shown in Figure 2.9.

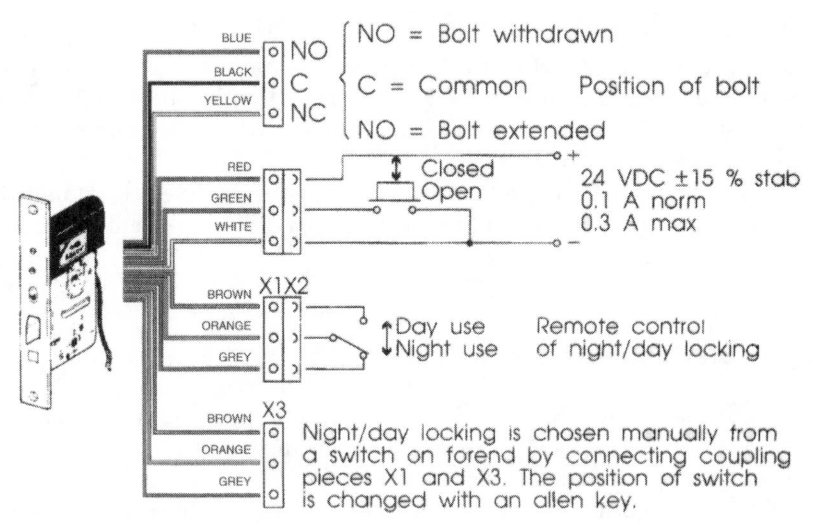

Figure 2.9 *Wiring Day/Night function*

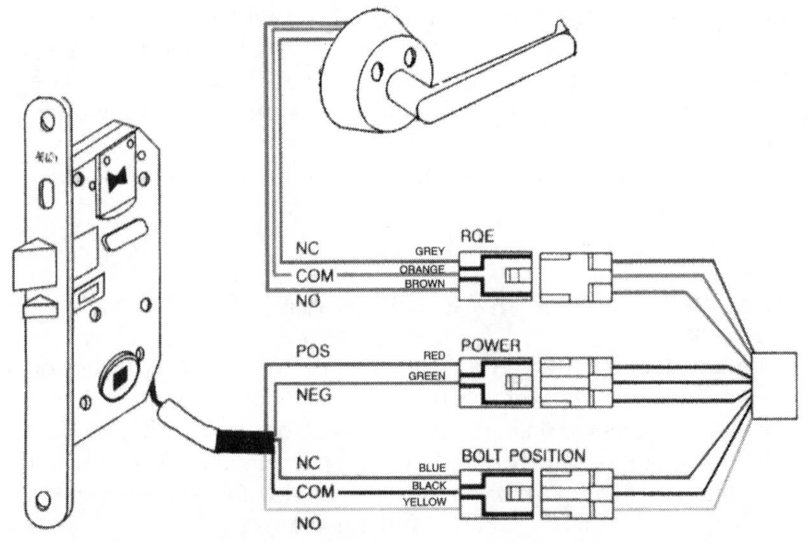

Figure 2.10 *Wiring solenoid lock with bolt position monitoring and RQE inside handle*

Example 3 This example is to illustrate the typical wiring of a solenoid lock with bolt position and RQE monitoring of the inside lever handle. An integral microswitch indicates the position of the bolt. It should be remembered that the solenoid lock is one in which a power source energizes a solenoid to create a magnetic force capable of operating the dead bolt (Figure 2.10).

We can therefore conclude that with unmonitored locks the status of the lock and door is not compared by the access control unit, but with monitored systems a comparison is made so that notification can be given that a system has been violated. This can be a door forced open or left open beyond a specific time.

With networked systems alarm events can be viewed alongside standard access control events at a central monitoring point. This is a subject that we investigate later.

2.4 Discussion points

To fulfil electronic access control duty criteria, a door must have an electrically energized lock or release, monitoring equipment for the door, a door closing mechanism and control circuits.

The vast majority of purely mechanical locks do have an equivalent standard electrified version but consideration must be given to the provision of cabling, strength of the electric lock and how the replacement will affect the emergency exiting and any other building or security measures that the original lock satisfies. In addition there are aesthetic values to uphold, abuse by users and the state that the lock and door will resort to in the event of a power failure or systems breakdown or lock failure.

Discussions must balance all the needs and regulations to ensure that none are compromised.

2.5 Glossary of terms. Locking devices

Automatic deadlocking A device incorporated within a mortice or rim lock which locks the bolt or latch when the door is closed. This protects against end pressure being applied by such objects as celluloid or metal strip. The bolt or latch can only be retracted by the key or knob.

Backplate A small plate applied to the inside of a door through which the cylinder connecting screws and bar are passed.

Backset The horizontal distance from a door edge to the centre of the keyhole, keyway or follower.

Bolt throw The distance the bolt protrudes from the face of the lock.

Case The housing or shell that holds the mechanism.

Claw bolt A type of dead bolt which when locked has two metal claws which move outwards and sideways to secure themselves behind the strike plate. They give good resistance to jemmying.

Cylinder lock An alternative name for a lock which generally uses a disc mechanism.

Dead bolt An unsprung bolt which is controlled by a key only. An exception is in bathrooms and in escape lock applications where the dead bolt can be operated by a turn knob or lever handle from the inside.

Deadlock Usually a mortice lock which only has a dead bolt controlled by a key.

Door stile The vertical section of a glazed or panelled door which may determine the maximum backset or case size of the lock to be fitted.

Electromagnetic lock A means of providing locking by magnetic force using an electromagnet and locking armature.

Fail locked/fail secure In the event of a power failure or breakdown the door will be locked.

Fail unlocked/fail safe In the event of a power failure or breakdown the door will be free of fastenings.

Follower A normally square hole visible when viewing a lock case from the side. The insertion of a spindle, knob or lever handle will operate the spring or latch bolt.

Latch A bevelled device normally of metal that can be withdrawn by handle or key to open or close a door.

Latch bolt A spring-loaded bolt, bevelled on its striking face which operates automatically when the door is closed, or withdrawn by the knob, handle or key. It may also be referred to as a spring latch.

Lock monitoring Indicates that the bolt of the lock has been thrown.

Mortice A slot or hole cut into the edge of a door into which a lock or latch is fitted.

Mortice lock A lock designed for slotting into a prepared hole cut into the edge of a door.

Motor lock A lock through which a continuous power source energizes a motor controlling the main bolt.

RQE Request to exit function. The lock or egress handle is provided with a microswitch to shunt a door.

Rebate The rectangular step-shaped recess cut down the leading edge of a door.

Reversible lock A lock in which the latch bolt can be turned over so to adapt it to doors of either hand and opening in or out.

Rim lock A latch type lock often referred to as a nightlatch. It utilizes a spring bolt operated from the outside by a key and from the inside by a handle or thumbturn.

Sash lock An upright mortice lock with both a latch and dead bolt function.

Solenoid lock A lock through which a power source energizes a solenoid thereby creating a magnetic force to control the operation of the bolt.

Spindle The square steel bar located in a knob or lever handle and passing through the lock follower. It enables the latch and/or bolt to be withdrawn when the knob or handle is turned.

Striking plate A metal plate morticed and screwed to a door frame. Generally supplied with mortice locks and deadlocks, plus rim locks where an outward opening door application exists.

Throw The distance that a bolt travels from the unlocked to the locked position. Certain locks may have a double bolt throw activated by an extra turn of the key.

Thumbturn A device that operates like a handle or knob enabling the withdrawal or throwing of a latch or bolt.

3 Tokens and readers

In this chapter we consider the identification methods in common use and which are used to discriminate between authorized and unauthorized personnel before access to a given point is allowed. These methods extend from keypad entry systems through to the various access card, token and tag types and include non-contact identification and personal characteristic or biometric systems.

The identification techniques are presented to an electronic reader and then automated into the access control system. These techniques differ depending upon the application and level of security required.

This part is referred to as the identification stage because the decision made at this point by the reader determines the action that the system will take, be it to grant access or to generate an alarm condition. Entry can be granted on the basis of knowledge, in that the user is allowed entry because of what they know. It may be granted on the basis of possession because the user carries a token or tag that is regarded as authentic at that period of time. The other condition of entry is that the user has a physical attribute that the reader accepts.

When higher levels of security are specified the different techniques can be amalgamated so that two or more different conditions must be satisfied simultaneously in order that authorization is granted.

In addition to the different methods of identification that are in common use, we consider the advantages of each together with the number of combinations plus the levels of reject and accept that can be expected. There are other considerations that must be addressed to gain general acceptance by the users in order that the system cannot be abused. In addition there must exist a procedure to issue temporary credentials for casual staff or callers to enter a prescribed area.

3.1 Methods of identification

In access control we rely upon three fundamental forms of identification:

- Knowledge. Derived from a code or PIN being entered into a digital keypad. These form the keypad entry system technique.

- Possession. Based on the holding of a mechanical key which has a code held within its construction either in its notches, grooves or dimples or by data being encoded into a card or token. This method also covers non-contact identification.
- Personal characteristic. Referred to as biometric systems these verify the unique personal features of the person attempting to gain access and are used when high levels of security are required.

Single identification methods are, in the main, secure; however, when extra levels of security are needed combination techniques can be adopted in that a number of different technology systems must be satisfied simultaneously or within a prescribed time limit before access is granted. The most common of these is card plus code where the person attempting to gain entry must hold a valid card and have an authorized PIN. Therefore the potential entrant has possession and knowledge. In practice any system can be used as a dual approach method.

It is apparent that all identification methods have advantages and disadvantages and in practice they also have inherent levels of false accept and false reject:

False accept the admission of unauthorized personnel.
False reject the rejection of authorized personnel.

With dual approach techniques there will always be a higher level of false reject and a lower level of false accept. Code identification systems have low levels of false accept and false reject but these factors are attributed to the user rather than to the equipment. Knowledge-based systems are therefore inherently reliable in themselves but must trust the user to address the problems of spying or casual observance being employed by others.

Possession-based systems are governed by the card, key or token becoming damaged or stolen. In other respects they are both reliable and durable.

Personal characteristic or biometric systems are more complex. They were developed to negate the problems caused by the duplicating of tokens and unauthorized access being gained to PINs. Although biometrics are highly secure they have higher levels of false reject and this percentage value must be set in the equipment to keep the level to an acceptable minimum. The more onerous the acceptance criteria the greater the chances of the rejection of valid entrants whilst lowering the acceptance criteria may lead to false accept.

3.2 Code identification systems

Extremely versatile, the keypad is durable and inexpensive. It operates by validating the presentation of a code entered into its keys. The code is generally made up from a number of four to ten digits with errors or deleting being made by the use of additional keys such as * or #. The keypad may also have additional keys identified by letters of the alphabet for special functions.

The coded keypad is easily understood and not difficult to maintain as it features little hardware as a single technology, although it will also be found combined with other technologies to provide a dual approach.

The security of the keypad and system is very much related to the number of keys that make up the unit and the number of digits that form the code. A realistic level of security is offered by having 10 000 possible code combinations. It is possible to overview the number of combinations of a system by reference to Table 3.1.

Table 3.1 *Keypad code combinations*

No. of keys	*No. of digits in code*	*Combinations*
10	1	10
10	2	100
10	3	1000
10	4	10 000
10	5	100 000
10	6	1 000 000
20	1	20
20	2	400
20	3	8000

Example 10 digit keypad. 5 digits in code, i.e. 10 × 10 × 10 × 10 × 10 = 100 000
20 digit keypad. 3 digits in code i.e. 20 × 20 × 20 = 8000

In the majority of cases there are 10 keys, i.e. 1–9 plus 0, but higher security can be offered by a keypad with 20 keys, although it is normal to use a greater number of digits in the code in preference to doubling up the keys.

There are a number of methods employed with coded keypads to increase the difficulty of unauthorized persons trying to gain access through them. Keypads with scramble functions address the problem of spying by randomly changing the key top labels. These units come with a preview window showing which keys apply to the different numbers so that the finger patterns of entering the code into the keypad regularly change. Alternatively warning lamps or audible sounders can be linked to the keypad to deter persons who may attempt to enter multiple numbers at random in order to search for a code. These warning devices can then signal a security area or a reception point if a number of incorrect transactions are made. Equally a local sounder is normally adequate to stop an unauthorized person tampering with a keypad as it brings attention to the area.

Timers may also be incorporated in the keypad to extend the time taken for the keypad to respond if an incorrect code is entered. This method effectively closes down the system on each occasion that a wrong digit is entered and therefore extends the time that an unauthorized person must spend attempting to cheat the system or find a PIN.

Other considerations are that for security to be maintained, keys are to be regularly cleaned so that the keys used in a PIN and those keys that are unused are not identifiable as such by their appearance. Code changing should always be performed on a rotating basis and this will be found to be programmable at the keypad or selected by the use of jumpers or links. It is to be understood that certain code systems will have one code only for all users whilst others may have a multitude of different numbers interfaced with a central computer. However, if there are a great deal of PIN codes in use these tend to be combined with a card system, otherwise the security of the access control is seriously reduced because of the volume of valid codes that exist. This combined card system is an option to using codes and keypads featuring a greater number of digits.

Although in theory any other technique can be used with the coded keypad the most used method is the card system. In this type of control a common pass code has to be entered into the reader before the card is used. The card is then used to identify the individual with the reader linked to a data gathering system.

It will be found that the variety of keypads and options is enormous, and range from basic low cost units with simple functions to vandal resistant devices with high levels of environmental protection. The main advantages of the coded keypad are:

- Compact and easily understood.
- Different codes may be used to give access to different points and doors.
- Maintenance is easy.
- Not expensive. Easily replaced or repaired. Little complex hardware is needed.
- No cards or tokens need be carried so there is nothing to lose or forget.

- Can be supplemented with alarm and timer monitoring.
- Can be combined with other techniques following a common code being entered.
- Reliable.
- Programming is straightforward.

The disadvantages are mainly governed by the user being vigilant in not allowing the code to be viewed by others. Hands free is also not an option with the coded keypad.

3.3 Card-based identification systems

The card-based identification system relies on the reading of a token that may be a card, key or tag. There are a number of technologies that govern the levels of security offered and the durability afforded by the system. These items are encoded (have their information stored on them) by special manufacturing equipment and they must have an acceptable level of resistance to copying or the transfer of this information onto other cards. The resistance to the problem of copying and the physical resilience of the card itself vary between the different technologies.

Magnetic stripe (magstripe) cards

The magnetic stripe card is in fact the most widely used card, especially in the banking sector where it is used for credit purposes and bank automatic teller machines. They are inexpensive, easily manufactured and encoded, and can carry alphanumeric data. They are also the most common access system card and consist of a magnetically sensitive oxide strip fused onto the surface of a PVC material.

The card is encoded by means of bars of magnetized and unmagnetized material on the strip forming the binary code. This is achieved by changing the orientation of the north and south poles of the magnetic materials and using the transition distances between the poles to store the data. The magnetized and non-magnetized areas are represented by the numbers 1 and 0 to form the appropriate binary code which is deciphered by the reader on presentation. They are manufactured as standard or high coercivity, that is a measure of their ability to resist corruption by stray magnetic fields, with high coercivity cards being almost incorruptible. Coercivity force ratings indicate the strength of the actual magnetic field that is needed to erase magnetic materials and range from 300 oersted for low coercivity through to 4000 oersted for high coercivity. The durability of the card can be further increased by manufacturing them from a base material of polyester/mylar laminate in preference to PVC.

The magstripe card is available in many different formats so is able to contain varying amounts of information and in operation it is extremely reliable so error rates in readers are low. International Standards Organization (ISO) and American National Standards Institute (ANSI) data formats do exist but these are often found to be different to those offered by equipment manufacturers.

The magnetic stripe card is disadvantaged in that it may be physically damaged by misuse and its data can be affected by magnetic fields even when they are of only low potential. The other problems with this form of card are related to the high volume of equipment available for the reading and copying of cards so that unauthorized duplication and copying can never be entirely negated. However, the magstripe card does continue to command a widespread use for mid-security access control applications. It can also be supplemented with an additional and different card technology to improve its security against unauthorized duplication.

In operation the magnetic stripe information is read by swiping it through a reader or by inserting it into position in a slot. Insertion type readers are not easily affected by environmental conditions and are the most suitable for external duty. They also shield the sensitive reading heads within the body of the housing, although all magstripe readers do suffer if used in onerous exposed conditions. Certain motorized insertion readers can regulate the speed in which the card passes through the reading head and therefore increase reading accuracy.

It will be understood that the magnetic field from the card is not strong so the quality of information transfer does depend on the strength of the magnetic properties in the stripe and the cleanness of the card reading head.

These cards can be colour coded so that the card is only granted access when the indicator lights on the reading head correspond in colour with that of the card. The rule is simple: if, as an example, the green indicator is lit on the reader only green user cards can open the door and other colour cards are currently invalid. This is governed by time zones that switch between colours so that only certain colour cards can be swiped to gain access during normal working hours. Other colours are allowed at other hours and some colours may be authorized at all times.

It is possible to further individualize all magstripe cards by the printing of photo information onto them. Although providing a photo ID card is not difficult it must be manually identified at the point of access. The reading technology is therefore complex and expensive so it tends to become a second line of access control used to confirm an authorized entrant if challenged by security personnel once inside the protected area.

Barium ferrite (BaFe) cards

The role of the barium ferrite or magnetic spot card has largely been taken over by the magnetic stripe card but they are still found in some older systems. They comprise a laminate of three-ply form having the two outer layers made from PVC which enclose a barium ferrite rubber mixture that is capable of being magnetized. They resemble a solid PVC card and are encoded with their information by feeding them through special purpose equipment.

The term magnetic spot was derived from the format of the card that has small spots or beads of barium ferrite trapped in the centre ply and formed into a matrix display. When the card is passed through the special purpose equipment the magnetic orientation of all the spots can be influenced and arranged to produce different magnetic field patterns. The magnetic flux areas of the card have the barium ferrite central layer turned into miniature polarized magnets so that an array of circular magnetized zones are created. At a later stage the card can be newly encoded by subjecting it to a strong magnetic field that is capable of altering the magnetic orientation of all the spots.

Insertion readers are used to decipher the card that is carried out by the reader head sensing the position of the beads in the matrix and the magnetic orientation of them. In a similar fashion to the magnetic stripe card the magnetized areas and the non-magnetized areas of the card are represented by the numbers 1 and 0 when presented to the reader head. This data is formed into a binary code and this is then cross-referenced to form an alphanumeric code if the information is to be more than of a numeric nature.

BaFe cards are inexpensive and durable since the spots are embedded in the material, although they can be affected by stray magnetic fields. The cards give reasonable resistance to unauthorized duplication. They are not used alongside magstripe cards as concerns exist that the magnetic fields created by the BaFe card can corrupt the former type.

Magnetic watermark cards

These cards are similar to the magnetic stripe version and are used in medium levels of security for access control. The essential difference between that of the magnetic watermark card and the magnetic stripe unit is that the former is encoded with its data before the card is constructed so that the information is not later applied to the blank card's magnetic strip. In manufacture the watermark card has its magnetic strip, complete with the information it holds, held between the outer two layers of PVC which is then heat treated to form a card with the data-holding layer dispersed between the PVC layers. It therefore has a watermark that is not visible to the human eye. With the watermark technique of manufacture the code is permanently held in the card and cannot be changed as the strip is encoded when still in a wet state and before the magnetic coating is heated and sealed. These cards are therefore not capable of being copied but must be decoded by special purpose equipment.

The magnetic watermark card is durable and secure and the information that it holds is difficult to read or transfer. They do not scratch easily and are advocated for applications in which there are fewer card users of mid-security risk.

Wiegand cards

A high security card using magnetic field technology but with the field produced from within the reader by an electromagnet is sometimes referred to as the Wiegand effect. It is a patented technique that employs an array of wires or metallic strips that are embedded in the card in two clearly defined tracks. These wires can accept and store a magnetic charge within their outer sheath. They also have an inner core which is twisted and that can be magnetized but will not store the charge. Special metal alloys are used to create unique magnetic properties not normally attributed to common iron wires so that the core of the wire has a different magnetic property to its shell. This forms a condition called bistable magnetic action.

These wires, which are located at specific points, act as a signal generator when a particular magnetic field energizes the inner core but is not of adequate capacity to magnetize the outer wire skin. The two tracks of wire that will be found in two regular rows represent 1s and 0s in the binary code and it is these that are read by the head of the card reader.

The card reader has two essential components, namely the magnetic field generator and the reader head. As the wires that are contained within the card move through the magnetic field of the reader the card generates both an electromotive force and an electrical current and this is fed to the outer skin. The change from the absorbing of the energy to the transmission of it therefore creates the electrical current and signal that is interrogated by the reading head. It is this pulse that is sensed by a coil and determines if the signal is a 1 or a 0 as governed by its location within the card (Figure 3.1).

The Wiegand card has many advantages:

- The durability of the card is high as the wires are embedded and the unique nature of the technology makes the counterfeiting of the card extremely difficult as the supply of Wiegand wires is controlled by the manufacturers.

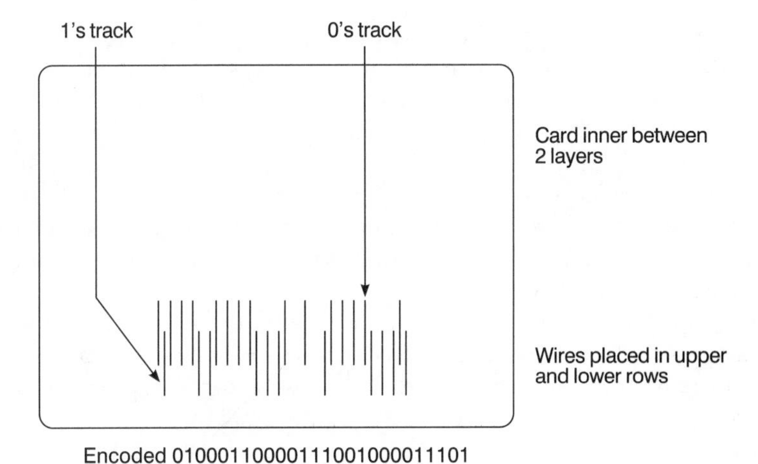

Figure 3.1 *Wiegand card*

- The card cannot be affected by electromagnetic interference or radio frequency interference.
- The energy released is of greater magnitude than that of many other magnetic-based cards so the card can be read without such close contact with the reader. This enables the head of the reader to be encapsulated and to give better protection when used in areas where the environment is more severe. These may be swipe or insertion readers.

A disadvantage of the Wiegand technique is that the cards are permanent and cannot be newly encoded.

Optical bar code cards

These are used in low risk security applications that generally have a high turnover of staff. Visibly bar codes appear as a set of parallel lines that vary in thickness to form a pattern of light and dark that is interrogated by an optical reader to form a code. These stripes represent the card information. Bar codes can be printed directly onto almost any card or indeed label or object and the reading equipment and software used to interrogate these are inexpensive and readily available. The codes may be easily duplicated by copying so their security level is low but as they are easy to create they do have a role to play in casual entry through access controlled systems. To this end they can be quickly produced through a PC and printer. They are also easily understood and cannot be corrupted by stray magnetic forces.

The accuracy offered is governed by the optical reader and although not often found in permanent access control systems in their own right they can be used alongside other technologies to provide increased information. If a measure of increased security is needed, the bar codes can be applied in a form not visible to the human eye so that special readers are needed to optically view the code.

Internationally there are a number of different bar code symbol sets and it is also possible to find multiple bar codes that can be scanned as one extended bar code.

The optical bar code technique is therefore of low security and although of reduced durability it is not expensive and can be used for certain temporary and casual entry systems and the supplementing of other technologies to provide additional information.

Smart cards

These are becoming more popular in access control systems and they hold great prospects for the future. They are of credit card size but have integral microprocessing and data storage. The smart card is identified by the metallic contacts that it displays on its outer surface as these make contact with a computer terminal when the card is inserted. Essentially the smart card is a credit card sized computer with the microcomputer, battery and communications electronics and circuitry held between the outer protective PVC sheets. It has capabilities similar to a PC that extend well beyond the needs of even the most advanced and complex access control system as it can store enormous amounts of personal data. It is therefore perceived as a multilevel card with the facility to hold information related to cash and credit transactions through to licences held by persons, qualifications and insurance certificates and medical history.

The smart card can be updated from a single computer source so that information can be transferred to it direct from stored data.

The present generation of smart cards are not as durable as most of the other types of card currently in use and they should always be stored flat and its metallic contacts must be kept clean.

We expect to see great advances in this technology but currently there are three basic types: memory only, memory with certain hard-wired logic or microcomputer. The microcomputer also contains the microprocessor, non-volatile memory (NVM), random access memory and its read only memory which holds the data on the access control operating system.

The smart card is presently used to great effect in the banking sector as it is extremely difficult to copy and is reprogrammable even though its durability is low and its cost high. Nevertheless its use will surely extend further into the access control spectrum as its technology develops to make it somewhat more cost effective in security system applications.

Infrared cards

Sometimes called shadow cards these are cards backed in PVC but with special bar codes embedded into position. They use light sensitive infrared technology but with bar code principles for encoding. In appearance the card is opaque to natural light but once subject to infrared light the bar code casts a shadow behind the card and this is deciphered by the head of the reading equipment. The card can actually exhibit different colour contrasts and shadows by employing different levels or thicknesses of PVC over the bar codes so that the light will penetrate both the PVC and bar codes to different degrees. This makes copying extremely difficult as the bar pattern is not visible to the human eye and ensures that this technique can be used in high security risk applications.

The infrared card relies on the quality of the light sensitive reader head to ensure a good performance criteria and is not affected by stray magnetic fields because of the nature of its technology. They are, however, more specialized and reasonably expensive.

Hollerith and holographic/optical ID cards

The card systems that have been covered so far are in international use but the hollerith and holographic ID cards are more specialized and for this reason used only in limited applications, nevertheless they should still be understood. They both use optical techniques for their operation.

The hollerith card has a series of holes punched in it and is read optically to decipher the pattern of light transmitted through it. Alternatively it can be interrogated electronically for electrical current that is able to pass through it. Both versions are low security as they are easily copied but they can be changed quickly and are used for control of areas where the turnover of people and pedestrian traffic is rapid.

Holographic ID cards are at times called optical cards. These are a highly secure variant of the bar code card. They are manufactured from two layers of PVC with the information engraved onto the lower layer by a laser beam that burns minute pits onto its surface. This layer or skin is then covered by the upper layer that acts as a lens and also physically protects the data. The card effectively holds a hologram that transmits and reflects extremely fine beams of highly concentrated laser light in a clearly defined pattern creating a picture that is analysed by the card reader for reflection which varies in accordance with the data that has been etched in position. The information is therefore transmitted in a form of light conveyed in light patterns or holograms using a holographic generator technique.

These cards can carry an enormous amount of data, although they must be encoded at the point of manufacture. Forgery is difficult so they can be used in high security installations but are more expensive than many other card types.

Often the holographic card also holds a passport size photograph of the user with additional data that can be shown to security personnel if the user is challenged. Image databases on a computer also enable verification to be quickly made.

Mixed technology/dual technology cards

This is a technology that combines a variety of techniques on one card and provides different functions by the employment of these different techniques. The most popular combination is magnetic stripe in conjunction with Wiegand. The user may want to reduce the number of cards that they are committed to carry and the combining of the technologies on one card addresses this need as customer requirements often ask for different data to be held; technically this may be best achieved by using different technologies. In this respect bar codes are ideal for storing information on record keeping whereas magnetic stripe is efficient at accessing debit systems and Wiegand gives a high degree of security for gaining entry to designated areas. In addition the use of custom designs and photos on cards helps identify persons.

Consideration must be given to the physical size of the card and at the present time there are two standards that have been pioneered by the banking industry.

- CR80. A common credit card size. Fits all swipe and insertion readers.
  2.125 inches high × 3.375 inches wide × 0.03 inches thick.
  ISO form.
- CR60. A more square card than the CR80. Restricted mainly to swipe readers.
  2.375 inches high × 3.25 inches wide × 0.03 inches thick.

As with all system management methods the authorized card holder must have an honest approach to the electronic access control function otherwise the full system is at risk. To this end photo identification is ideal as it always confirms that the card holder is authentic. In the set-up and management role of the access control system there are certain procedures that can be recounted.

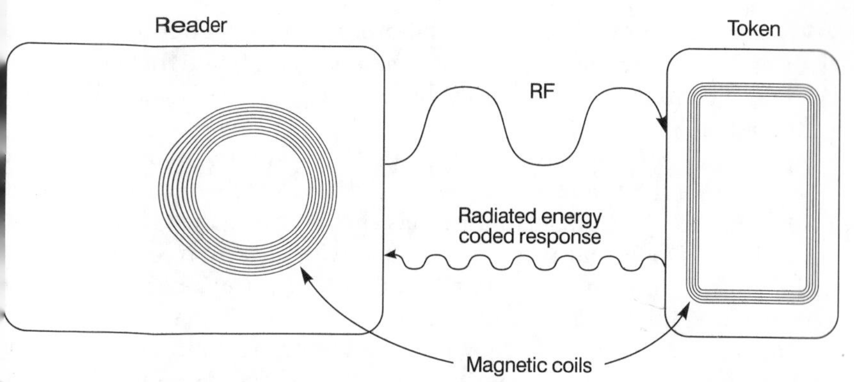

Figure 3.2 *Passive tuned circuit*

– Field powered circuits. These are more complex and tend to have a memory function. They have a power supply that is able to convert the radio frequency energy as emitted by the tag into electrical energy that is subsequently used to power the tag.

- Active tags. The tags carry their own onboard power source which is essentially a miniature battery. They are therefore active since they do of themselves provide power to the integral electronic circuits of the tag or token that contains the receive/transmit coils and memory logic PCB. The batteries become less powerful over time and need systematic replacement. These tags, however, are not governed by the strength of the reader's radio frequency field in the same way as the passive token and will generally have a range of up to 2 metres, although more specialized equipment will claim far greater distances.

'assive tags are smaller and lighter than active tags. They are less xpensive and have virtually an unlimited operating life, but active tags ave a greater distance reading range and can be scanned faster so are etter employed if persons or traffic are attempting to pass through a ontrolled area simultaneously. All transponders are available in different ›rmats that can be Read Only, Write Once Read Many, and Read/ /rite:

Read Only. These are encoded at the point of manufacture and cannot be reprogrammed.

Write Once Read Many. These are normally also encoded at the place of manufacture but alternatively can be encoded at the location of a registered user to avoid delays with data encoding. Otherwise they are the same as Read Only.

Enrolment

The entering of information or data regarding the level of access for users. Basic entries are:

- Name.
- Gender
- Address.
- Vehicle particulars.
- Employment function on site/department
- Telephone number/extension.
- Time zones on site.
- Levels of access.
- Identification of doors and areas to which access is authorized.
- Health and previous employment records.
- References of character.

Validity

This is the verification that the cards and credentials are current and have not been altered. These should all be updated and reissued on a rota. Termination should be considered so that non-active cards can be immediately retrieved to stop attempts at forgery.

Overview of card types

It now becomes possible to overview the different card technologies and the main advantages and disadvantages inherent in the various techniques.

The different card types that are currently in common use are shown in Table 3.2. This gives essential data on the technology, durability of use, resistance to forgery, ability to re-programme and an indication of the cost parish.

3.4 Non-contact identification systems

In non-contact identification systems the card or token does not need to touch a reader in order for the transaction to be validated. There is therefore no wear on the card or tag since no physical contact between it and the reader is made, so the non-contact identification card has an inherently longer life than that of the swipe or insertion device. These type of systems may equally be referred to as proximity systems as the tag only needs to be in the same proximity as the reader to be recognized using radio frequency transmissions.

Table 3.2 *Card types*

Card type	*Technology*	*Durability*	*Resistance to forgery*	*Re-programme*	*Cost*
Magnetic stripe	magnetic media stripe	M	M	Y	L
BaFe	magnetic pattern	H	M	Y	L
Magnetic watermark	manufacturing stage magnetic code	M	H	N	L
Wiegand	magnetic pattern	H	H	N	M
Optical bar code	light/dark patterns	L	L	N	L
Smart	microcomputer	L	H	Y	H
Infra red	light/dark patterns	M	H	N	H
Hollerith	hole patterns	L	L	N	L
Holographic ID	laser scan of holograms	M	H	N	H
Mixed technology Dual technology	combined technologies	-	-	-	-

Note: M – medium
L – low
N – no
Y – yes

The most popular mixed technology is magnetic stripe with Wiegand.
BaFe should not be mixed with the other magnetic media/patterns or microcomputer technologies to give a dual technology.
Magnetic stripe and magnetic watermark should also be isolated from each other.
For mixed technology the durability, resistance to forgery, ability to re-programme and the cost are dependent on the techniques that are combined on the card.

The readers can be held in vandal-proof locations and b
they are secure in operation. The readers operate with
consumption to the same levels as those of magnetic stripe a
card readers.

As there is no direct contact the speed of human or v
through an access point can be increased and cards or to
held in pockets or in display positions that enable hands-f
be achieved. Doors and barriers can open without any nee
swipe a card through a reader, although the reading rang
differ and with some it is still necessary to hold the card o
the reader but without having to make physical contac
proximity and hands-free are used for many different prod
it is advisable only to class products as hands free when th
read the tag without the user having to hold the tag
reader.

Proximity systems refer to a technique in which the tag
without contact being made with the reader but in pract
range may be less than half a metre. Clearly these sys
enable persons to pass through an area when both
occupied.

These systems may also be called radio frequency i
systems (RFIDs) and use the term transponder in preferen
tag. These transponders are used in conjunction with a r
that scans them for the data that they contain and this
processed by a computer. In practice the system has a
receiver unit which excites the magnetic coil in the card
range. This range is extended when the card also holds
battery. The coil in the card then generates a magnetic
represents a code within its memory chip. The transmitter
collects the magnetic pattern that it amplifies and tran
processor for analysis. There are essentially two types of sy
namely passive and active:

- Passive tags. The operating power is received from
generated by the reader. The tag is therefore passive in t
it does not generate energy, and relies on the strength of t
the reader and transmitter/receiver control unit for the p
to make them responsive. This is performed by the use of t
powered circuits:
– Tuned circuits. Within the tag an electrical circuit reso
enters the radio frequency field that has been gener
transmitter/receiver control unit and as it resonates it
energy back to the reader but at a different frequency a
coded response (Figure 3.2).

- Read/Write. These are flexible in use and allow encoding by the manufacturer or user and can also be reprogrammed to offer new levels of data and changes to the information they hold. This format is adopted where information is required by systems that are not interconnected and do not therefore share database information.

A further variable of the tag is the frequency at which the system operates. The system performance can be altered by a change to the size of the tag and the size and power of the reader, but the following frequencies are the most common:

- Low frequency tags operate in the range 100 to 500 kHz. They have a reading range of 10 to 200 cm but a low reading speed. They are of relatively low cost.
- High frequency tags operate in the range 10 to 15 MHz. They have a short reading range of 1 metre maximum but have a high reading speed. They are of medium cost.
- Ultra high frequency tags operate in the range 850 to 950 MHz or 2.4 to 5 GHz. They have a long reading range in excess of 10 metres with active tags and have a high reading speed. They are of high cost.

All these non-contact identification systems are high security but tend to be most used in the larger establishment, although they do offer rapid door operation for authorized persons and rapid barrier operation for vehicles. The hands-free concept is invaluable if the transporting of packages is a frequent event as doors can automatically swing open to allow access.

A major consideration of these systems is the location of the reader and its operational function as it must both stimulate the token and then analyse the data that is refocused back to it. The reader or transmitter receiver must hold the power source and the antenna plus the processing electronics. These may be held as an integrated unit in one housing or they may be separately located. The antenna can be found in one of many forms to the point of being concealed and encircling the full perimeter of a door.

All these systems, however, because they use radio signalling technology, are affected if the reading equipment is in close proximity to other sources of radio frequency energy emitted by televisions, computers, printers and such. The signals are also weakened by local metallic materials, and metal door frames are a distinct hazard.

The non-contact identification system can use a multiple of codes and the tokens are manufactured and tuned to re-radiate a prescribed number of frequencies, although a greater level of accuracy is achieved if tuned circuits that operate at adjacent frequencies are not used within the same token.

The passive tag can also at times be found alongside other technologies such as magnetic stripe to carry additional access data for logging purposes.

As an overview we can say that the non-contact identification system is in the main used to achieve convenience through ease of access but coupled with high security. These techniques also avoid vandalism being carried out on conventional readers or abuse of them by inconsiderate users. Radio frequency readers may also be mounted behind protected partitions made from standard building materials yet still carry out their reading function. Vehicle recognition is easily accomplished and more powerful equipment can be adopted to drive active tags that can be read over extensive distances.

With respect to the criteria of Table 3.2 we can endorse the capabilities of non-contact identification.

Technology:	Radio frequency transmission.
Durability:	High.
Resistance to forgery:	High.
Reprogramme:	Depends on system type.
Cost:	Medium to high.

3.5 Personal characteristic systems

Also called biometrics, as derived from the term biological data, they are used when the need for high security access authorization offsets the relatively high cost of the technology. These systems are based on the use of physiological patterns that are interpreted by a number of different techniques. They were developed in order to defeat the problem of cards, tags or tokens being shared, lost or stolen or PINs being observed.

Although personal characteristic systems do create extensive computer data and files they can enrol individuals in a matter of minutes and verify the authenticity in only a few seconds with high levels of accuracy. Complex systems can update individual templates, which hold the information, on a daily basis to take account of the ageing process. There are a significant number of personal characteristic systems in use and certain of these also incorporate a PIN for extra high security, but of course this does lead to a slower verification process. Using advanced data compression techniques certain systems can be found to offer biometric data such as fingerprint identification on cards and although this leads to easier storage of computer data the cost of reading equipment can be excessive.

Certain aspects of our physical or behavioural characteristics are unique to each individual and they form the basis of this technology. There are particular advantages and disadvantages:

- Advantages. Gives automated verification that the person attempting to gain entry is authentic. No other systems can provide such verification. No additional means of entry is needed other than the unique personal trait.
- Disadvantages. Cost is high. The length of time to verify is longer than that of most other systems and levels of accuracy must be carefully calibrated.

The stages in the process of implementation are enrolment, template creation and comparison:

- Enrolment. The point at which the relevant information is stored.
- Template creation. The base for comparison of every subsequent entry.
- Comparison. The acceptance or rejection criteria. Readings of biometric features produce an algorithm to determine the accuracy of the comparison. The client with the installer selects a level at which acceptance is wanted according to his type:

  Type 1. False reject
  ⟶ criteria
  Type 2. False accept

  Type 1. False reject.
  This is the degree of rejection and a low threshold can be set so that only one in 100 000 authorized personnel will be denied entry.

  Type 2. False accept.
  This is the case in which a non-authorized person is granted access. For high security a high threshold is selected. This prevents unauthorized entry but means that a greater incidence of genuine personnel will be denied entry.

The most popular biometric characteristics are: fingerprint/palm recognition, hand recognition, eye pattern identification, voice recognition, keystroke dynamics, handwriting and combination techniques.

Fingerprint/palm recognition

This may be applied to the finger only or also to the palm print. In practice the finger or palm is measured on a minutiae-based formula or by pattern recognition that digitizes the pattern and then seeks an image match. Records of the patterns of the finger or palm print formed by its loops, whorls and geometric shapes are assessed against the template, whilst in the case of minutiae techniques the smaller scale details are verified by looking at the endings and branch points of the troughs and peaks. These points are mapped and linked with straight lines to achieve a matrix for grid mapping and that are held on the template.

The systems differ in the area scanned and in the methods used to gather the detail. This can be through the use of ultrasound or a light-based system and can extend also to the gathering of data of the blood vessel pattern. Multiple patterns ensure that a severed body part cannot be used to falsify identification as two different biometric readings must be taken simultaneously. As an example, whilst the fingerprint reading is being taken a blood/oxygen reading is also needed. Fingerprint systems are becoming quite popular as they are easily understood but the systems can be slightly disrupted if cuts or sores appear on fingers or if grease or such contaminates the fingers and the scanning plates.

Hand recognition

This technique assesses the geometry; height, width and distance between knuckle joints and finger length in an algorithm for comparison. Also termed hand geometry systems they map out the hand format by camera scanning techniques and this is digitized on disk. These systems are durable and easily understood but the reader is physically larger in size.

Hand recognition is reasonably accurate since the shape of all hands is unique, but they do tend to give higher false accept rates than fingerprint or palm recognition, although their rate of speed of verification is more rapid.

These systems can also be found combined with data features of blood vessels.

Eye pattern identification

Achieved by an iris scanner that emits a low level infrared beam through the pupil to obtain a square image of the capillaries at the back of the eye. The formulation of these capillaries on the retina is unique to all individuals. The scanning procedure is carried out by presenting the eye to an aperture. This is focused on a selected point and known as retinal scanning.

Other techniques can scan the surface of the eye so focusing does not need to be carried out and the presentation can be conducted at a short distance from the reader. This is called iris identification.

Problems can occur with persons wearing glasses, contact lenses or have light sensitive eyes or those who feel timid about their eye being assessed.

Voice recognition

Essentially this is a voice synthesizer technique to recognize speech characteristics. The system identifies the voice characteristics of a given phrase to that of one held in the template. The comparison takes account of air pressure and vibrations caused by the movement of the larynx. Spectral frequencies are adopted as these account for variations encountered in normal speech as attributed to colds and infections or such.

These are not generally performed as one function but access to the voice system reader is normally only given following the entry of a PIN or a different technique being satisfied first. The templates that hold the voice are regularly updated at the point of use to account for variations attributed to age. In relation to other biometric systems the voice recognition technique is less expensive. They are secure when part of a dual technology.

The voice synthesizer system must be carefully sited because the voice pattern is disturbed by background sounds. The scanners are neither large nor complex enabling their use in external applications if weatherproofed.

Keystroke dynamics

This technology measures the use of a keyboard being operated for stroke assessment. This must be done to a predetermined criteria. It measures pressure and speed of operation together with accuracy. It is somewhat specialized and not as popular as handwriting techniques.

Handwriting

This is the automated recognition of a signature or other writing based on a multiplicity of writing styles. The technique takes into account the pressure used to create a signature in addition to the form of the signature. These are found in two distinct forms. The first type needs a reception point to confirm acceptance of the signature via a VDU screen and is a presentation technique. The second form is classed as a dynamic technique because in the same way as keystroke dynamics it adopts a computer to verify the signature and also checks for the pressure used in writing out the signature.

These are only used in access control systems that are not subject to heavy pedestrian traffic because the procedure of verification is slow. It is often the case that a PIN is inserted into the system first so that the computer can then more easily identify the person seeking entry and search through the database in preparation for comparing the subsequent writing.

Combination techniques

Although in theory any techniques can be combined for verification purposes, the most used when biometric technology is adopted is card technology with fingerprint encryption. Verification in this case is against a template held on a hard disk.

There are certain more specialized technologies worthy of mention, namely body odour, facial recognition, random voice identification, height/weight measurements and blood vessel/bone format.

Body odour

This technology uses a chemical process to recognize odours but these systems are not particularly accurate. Body odour readers are not affected by perfume/scents as the compositions of these are different to those formulated by the human body.

Facial recognition

This technique compares an image held in a database to one scanned by a camera. Not popular but when automated is a more acceptable method than using manned guards to perform the verification.

Random voice identification

Used to increase the security of voice recognition by ensuring that a taped voice cannot be used to defeat it. It operates by requesting that the person attempting to gain entry recites a further phrase selected at random.

Height/weight measurements

This determines the presence of an individual in a particular location and proves that the individual is within given guidelines. It is not uniquely used to give access without extra verification.

Blood vessel/bone format

These systems are generally an extension of hand recognition systems but they also scan for image matches of blood vessels and bone structure using ultrasound.

Table 3.3 *Personal characteristic/biometric recognition*

Type	*Technology and considerations*
Fingerprint/palm recognition	Measures on a minutiae-based formula or by pattern recognition
Hand recognition	Assesses geometry, height, width and distance in an algorithm for comparison
Eye pattern identification	Either a square image of the retina capillaries or surface of the iris
Voice recognition	Speech characteristics using a voice synthesizer technique
Keystroke dynamics	Stroke operations assessed for speed, pressure, etc.
Handwriting	Signature or writing recognition presented via a VDU or by dynamic assessment
Combination techniques	Different access control technologies combined
Body odour	Recognition of human odours
Facial recognition	Camera scanning
Random voice recognition	Supplement to voice recognition with additional phrases entered
Height/weight measurement	Basic physical assessment only
Blood vessel/bone format	Supplement to other techniques but generally hand recognition

Overview of personal characteristic systems

At this stage we can overview the main types of personal characteristic traits or biometric patterns (Table 3.3).

It remains to say that biometric systems are not ideal in all situations and they are considerably more expensive than the majority of code, card, tag or token system. However, they can be used confidently where security measures must be high and human traffic flows need not be rapid because of the slower process time of the scanner.

Biometric readers often contain sensitive electronics and by their very nature they are often greater in size than more conventional readers so the installer must be aware of problems that could be attributed to vandalism or the weather. Consideration must also be given to persons who may be timid to biometric verification and to those groups with physical disabilities or with artificial limbs.

The personal identification system is nevertheless at an advantage in that it accurately and specifically identifies a person as an authorized entrant and does not apply this assessment to a card, code, token or such.

3.6 Discussion points

There are many techniques that the engineer can select from in order to find the best method of seeking the authorization of a potential entrant to gain entry to a protected area. The applications govern the selection of the most effective technology but there must be a willingness of users to accept the technique selected and not be intimidated by presenting their credentials for verification.

For entry to highly secure or sensitive areas biometric practices may be used but the reading technology is expensive in relation to other traditional forms of identification.

Keypads using PINs are convenient because possession of an object is not necessary, but unfortunately code security can be compromised by careless individuals.

Cards are unique to the holder and can hold other data. They are also easily administered and voided and can be colour coded for specific time zones. Proximity tokens are also convenient although active tags have a limited life because of their onboard battery.

In real terms there is a technology for every duty but the selection process must be thoroughly discussed to ensure the widespread acceptance and cost effectiveness of the chosen type before selection of the controller. The next stage is to interface the reader equipment with the controller that is the heart of the access control system. This is covered in the next chapter. However, before we go any further we need to discuss the function of a reader and token and how their operation is put into practice.

Figure 3.3 indicates the architecture of a coded keypad driven by a 12 volt supply that will operate a relay when an authorized code is selected and then switch this voltage also to an electric door release. A doorbell can be operated by pressing the additional 0 key and the keypad has a duress output through a further relay acting as a panic/ambush function and signalling a sounder if additional keys are depressed to the actual code.

This keypad could replace the mechanical lock in almost any situation but it needs additional functions to make up the comprehensive access control system covered in later chapters.

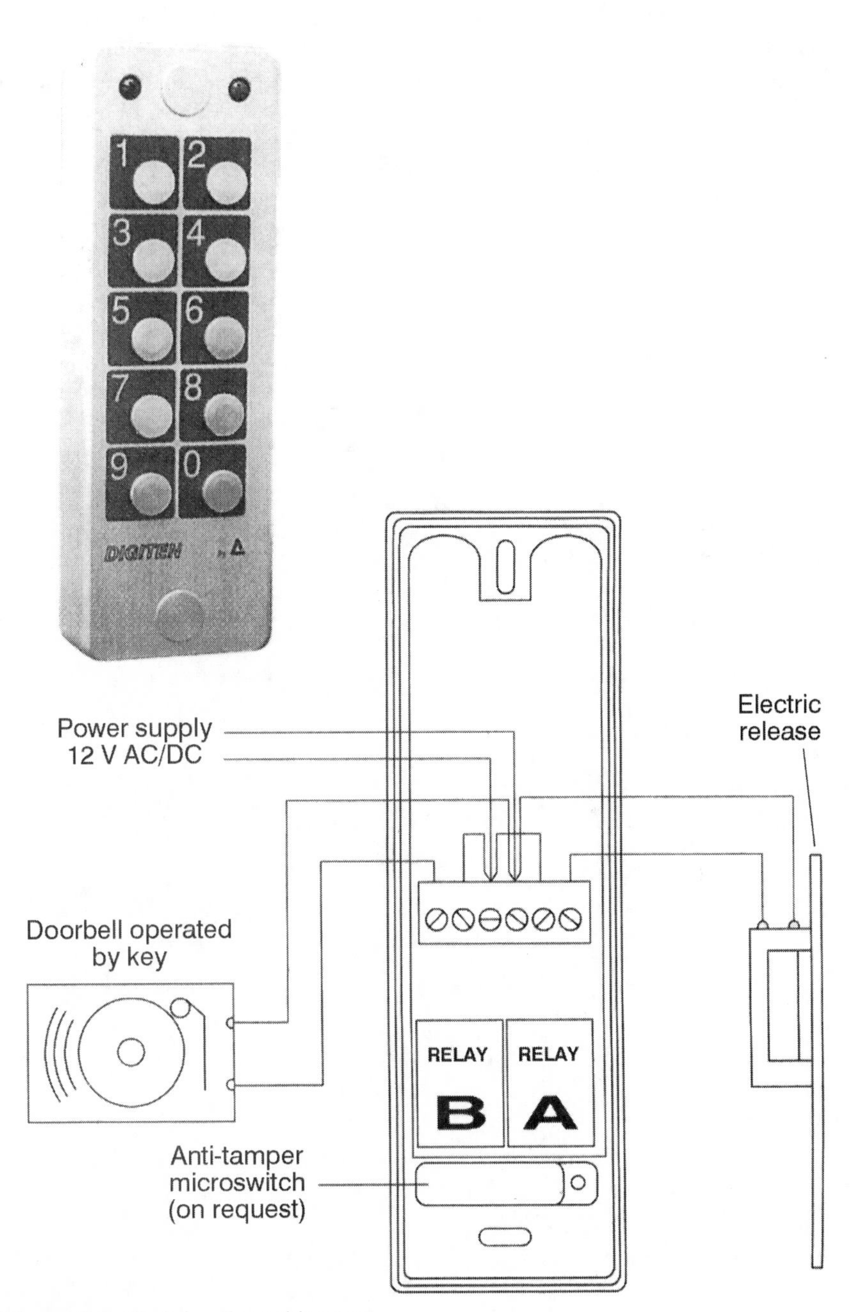

Figure 3.3 *Code/reader achitecture*

The keypad can be both weather resistant and anti-vandal using special anodized aluminium, high impact polymer mouldings or stainless steel housings and buttons. The printing can be indelible and the indicator LEDs pyrex protected. The protection levels apply equally to any reader and must be discussed in practical terms within the cost restraints of the system being specified.

4 Controllers and systems

In Chapters 1 and 2 we looked at the perimeter protection offered by electronic access control systems, which included barriers, gates and doors and physical restraints that prohibit entry beyond a given point unless authorization is granted. Chapter 3 introduced the methods of identification or the credentials that must be presented to the system reader or scanner for this authorization to be given. In practice access systems vary enormously so we arrive at the point when the functions that the system must perform are to be considered whilst recognizing that different sites must implement the controls in various ways. This involves many aspects and is influenced by the level of security that is needed. There may be different levels of access for the authorized persons who will use the system and these levels can change at different times of the day, week or night. Logging can also be implemented for attendance criteria to be retained and there can be automatic locking and unlocking together with integration of other security techniques and alarm outputs.

The central control or processing unit acts as a computer network and this has a multitude of features that can be added to the system programming. Particular software will ensure that the exact degree of security that is specified can be achieved under practical conditions and within the given cost restraints.

The ways of configuring the central control unit vary and the system may be compact so that all the electronics are held in the control unit itself. This configuration is less expensive to purchase, easier to install and is ideal if the doors are not under threat of attack. The electronics, however, if housed in a separate housing make up a more secure installation and these systems are more flexible in their ability to connect to different action points. These methods are all considered in this chapter.

There are a number of methods that are used to cable the systems and these techniques should be understood together with the various forms of cables that can be used for the specific function that they must undertake. These cables need physical protection afforded to them and must not be installed in such a way that they can have electrical interference induced into them. The concepts of dealing with this problem must be addressed. Allied to this is a need to understand the role of the CE Mark and electromagnetic compatibility and the restriction of interference emissions.

4.1 Single door controllers and software

In practice the most basic form of access control system is the simple door intercom that tends to be supplied as a kit. This is often referred to as a door entry system and does not fully satisfy the degree of operation and function offered by the full access techniques in which we are interested. Nevertheless there is an overlap in the industry so it is worth looking at the basic intercom as a starting point. It consists of a transformer or power supply, external door speech unit and internal telephone that may also have a video screen to give an image of the caller. An integrated door release facility enables the user within the premises to interrogate the caller from a remote point before granting access. The external door speech unit may also feature a reader enabling a caller to gain access without a need to contact a person within the premises if they can satisfy the reader with their credentials. The reader and controller will normally be integrated. It is a cost-effective but nevertheless low security system.

The system as described has a role to play in many applications and its security can be enhanced by purchasing the controller or processing unit as separate identities and mounting them in a remote and secure environment away from the reader. The reader is therefore the only part of the system on display and can be damaged by misuse, vandalism or by extreme weather conditions. This is apparent by reference to Figure 4.1 and illustrates how we progress from a simple intercom system into access control techniques.

Figure 4.1 depicts the most basic of architecture for simple techniques. The door entry system can be extended to include several telephones or stations and is easily installed. The power supply unit or transformer is placed in a protected indoor environment such as a cupboard and wall mounted onto a building material that is non-conductive. Connection is made to the mains supply by a dedicated spur.

The wiring from the low voltage AC or DC side of the transformer or power supply tends to be performed with twisted pair cabling with other similar cabling being taken to the additional components of the kit such as the door panel/speech unit and the lock release. The connections are clearly identified and options of trade button or time clocks can be added. These options are used in conjunction with each other to allow tradesmen access during particular hours only. The more advanced door entry or intercom system will also include privacy functions to block incoming calls and prevent monitoring and conference calls to initiate conference with a limited number of stations.

The true access control system, however, forms a much greater subject. In the first instance we should determine the functions that could be associated with a system that only controls one door before trying to

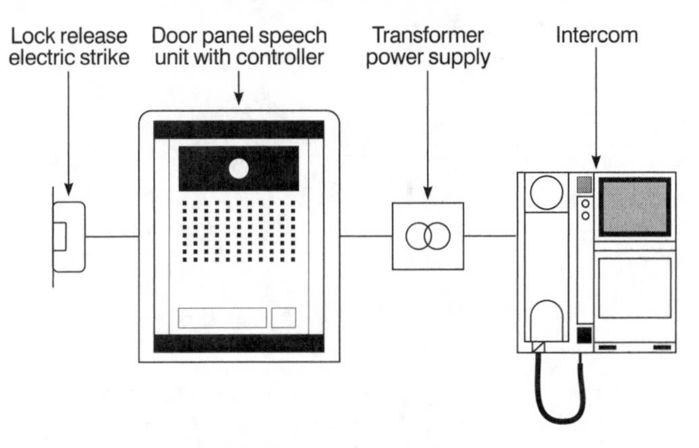

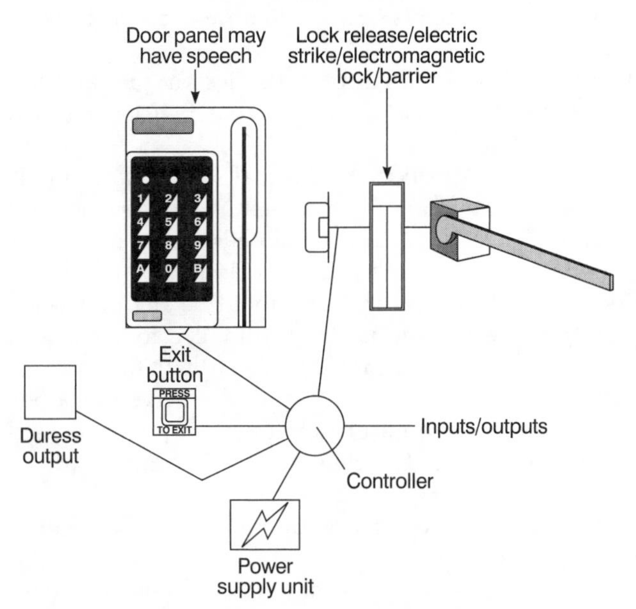

Figure 4.1 *Intercom/access control systems architecture*

extend our knowledge into multidoor systems. The equipment would consist of the following:

- Credentials. The identification method, which could be a code, card, tag or token or it may be a biometric characteristic. The reader or scanner will verify the transaction in conjunction with a processing unit.

- Door lock/bolt. This is controlled by the system to automatically lock the door/barrier in position.
- Processing unit. This may be integrated with the reader/scanner or mounted separately.
- Power supply. The device needed to drive all the equipment plus any ancillary components to include sounders and visual signalling parts.
- Interconnecting cables for control, power and communication signalling.

In addition to the foregoing the following may also form part of the system:

- Anti-passback. An arrangement where the software flags a card as either 'in' or 'out' when used at a specific reader. The anti-passback can be local or global, i.e. per controller or per system. It can also be timed and is used to stop the possibility of entrants passing their cards back to another person.
- Anti-tamper. An alarm fitted to the reader on mid- and high-security applications that can detect attempts to enter enclosures or subject it to high levels of shock.
- Door open sensor. An electromechanical switch is used in its simplest sense to monitor the door for being left open by changing the state of the switching contacts. A door-not-closed alarm can sound if the door unlock period is expired and is used with the door open timer. In high security applications this function should be able to recognize a door propped open, a door that has tried but failed to latch and a lock that has had an object inserted into it to cheat the latch bolt.
- Door open timer. Used in conjunction with the door open sensor it allows a measured amount of time to expire following a door being opened before an output is generated to inform of a door being left open.
- Door locked sensor. Integrated within the lock it is usually referred to as a monitored lock.
- Door lock timer. Controls the amount of time that the door lock remains powered once it has been energized to open or is overriden if the door closes before the time expires.
- Door closer. Physically closes the door once entry or exit has been made.
- Duress. A function that can give an alarm output to indicate that an unauthorized person is attempting to force entry through the controlled point by threatening an authorized person. This signalling can be achieved by the authorized person using a duress feature that could be achieved by entering a secret duress code or following a specific but non-standard procedure using a card or other entry technique.

- Exit (egress) button. Unlocks the door electrically from the inside by the pushing of a button to allow exit to be made.
- Remote lock activation. This is a means of activating and securing the lock from a remote position to stop any further entry being made in the event of an emergency, although emergency exiting is still allowed.
- Remote lock deactivation. In contrast to the remote lock activation technique this unlocks all doors so entry and exit can be made.
- Voiding of credentials. The method of voiding the codes, cards, tokens, tags or other credentials that are not any longer wanted by the system. It is the method of reprogramming the controller.
- Sequence analysis. The scanning by the system to locate credentials presented within very short time bands or the excessive use of the credentials to gain entry into specific areas. It can be compared against normal routines to gauge if any abuse of the system appears apparent and can be amalgamated with personnel tracking and attendance records.
- Alarm input. Triggered by a contact going open or closed to allow a further electrical function to be fulfilled or for the system to be linked to a different system that may also be related to security.
- Relay output. This is an output through the volt-free contacts of a relay changing over; it can also be used to drive ancillary equipment. This is generally for electrical equipment of a higher rating.

The interfacing of other devices or systems with the controller can be achieved by the alarm inputs or relay outputs but these must be checked for rating against the equipment they are to power or against the voltage and current they must be able to switch. An additional relay or contactor can always be specified for heavier duty as illustrated in Figure 4.2. The standard relay coil voltages are shown in Table 4.1 and the materials for the switching contacts are found in Table 4.2. See also Section 5.5 which overviews integrated systems.

At this stage we can refer to Figures 4.3 and 4.4 as examples since these show how a single door controller or processor can be configured.

Figure 4.3 is a standalone pushbutton coded access system with a prescribed number of user programmable codes held in a non-volatile memory (NVM). These can be entered as a string of numbers up to, for

Table 4.1 *Standard relay coil voltages*

AC coil	6 V	12 V	24 V	–	50 V	220/240 V
DC coil	6 V	12 V	24 V	48 V	-	220/240 V

Table 4.2 *Relay switching contacts materials*

Low load current ⟵						⟶ High load current
PGS alloy (platinum–gold–silver)	AgPd (silver–palladium)	Ag (silver)	AgCdO (silver–cadmium oxide)	AgNi (silver–nickle)	AgSnin (silver–tin–indium)	AgW (silver–tungsten)
High resistance to corrosion. Mainly used in minute current circuit	High resistance to corrosion and sulphur	Highest conductance and thermal conductance of all metals. Low contact resistance	High conductance and low contact resistance like Ag with excellent resistance to metal deposition	Rivals Ag in terms of conductance. Excellent resistance to arcing	Excellent resistance to metal deposition and wear	High hardness and melting point. Excellent resistance to arcing metal deposition and transfer, but high contact resistance and poor environmental durability

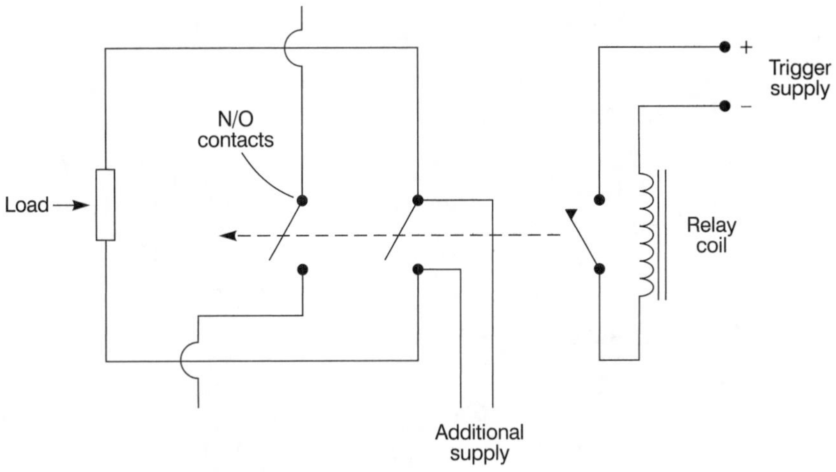

Figure 4.2 *Interfacing of access controller*

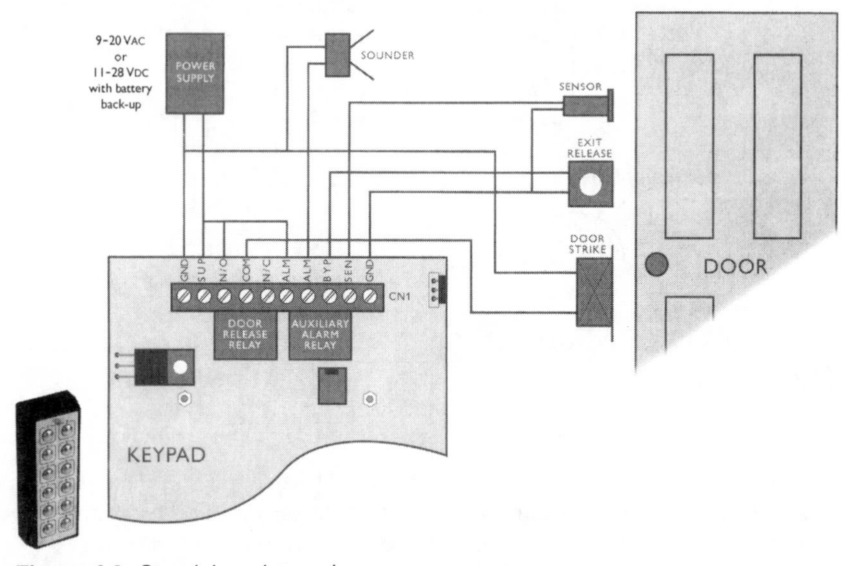

Figure 4.3 *Standalone keypad*

instance, twenty digits long to avoid discovery by an onlooker. Each code can be programmed to give a timed or latched release operation.

The unit has an LED indicator and an internal sounder for keypad feedback to confirm code acceptance or rejection. In addition to the door strike output the keypad offers a programmable auxiliary relay which can be used to indicate tamper, forced door, penalty count and door ajar via

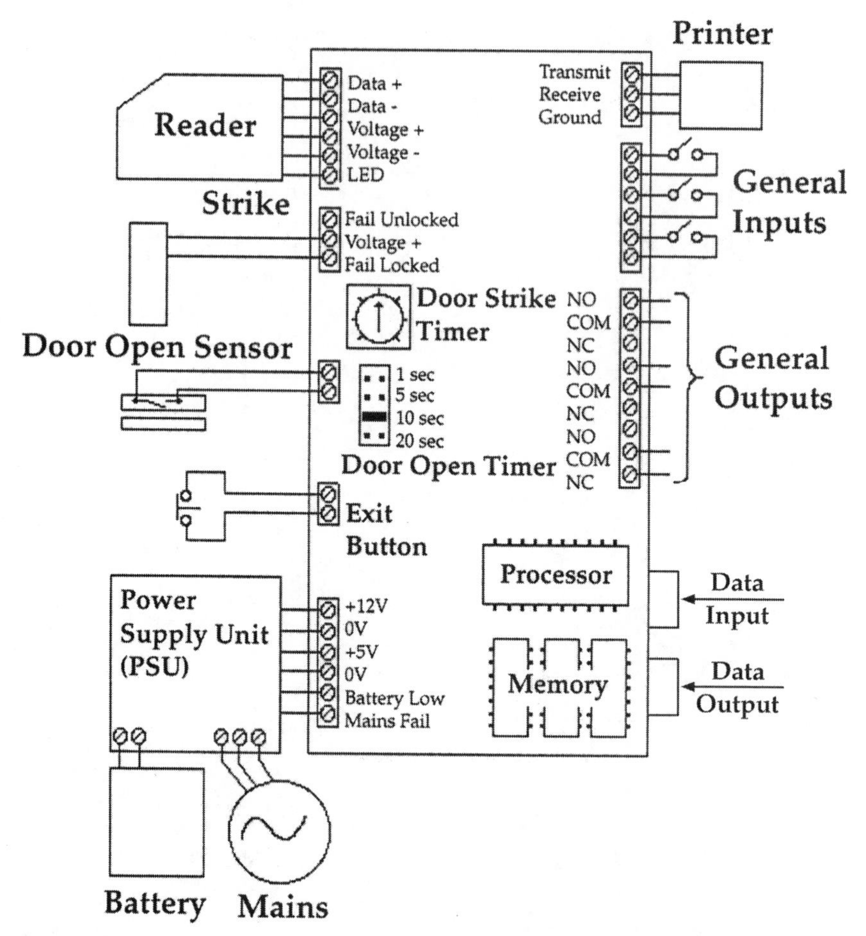

Figure 4.4 *Single door controller*

a remote sounder. The relay could otherwise be used to interface with an intruder alarm. Programming is via the pushbuttons. The programming mode is entered through a user defined master code.

The alarm outputs are available via the auxiliary relay contacts in descending order of priority as: tamper; forced door; penalty count and door held (ajar). The penalty count is a manufacturer's means of ensuring that no more than some twenty digits may be entered prior to a valid entry code. If this number is exceeded the keypad is disabled for a preset penalty time.

Figure 4.4 is an example of a further controller and it can be seen how the terminals are provided for connection of all the devices that can make up an essential system whilst the controller will also have integrated features of microprocessor and memory. The microprocessor has the memory contained within, alongside the information that determines the validity of any transaction requested.

The memory itself stores the data as to the means in which the controller must respond together with the information of the site and components of the system plus the historical data. A certain degree of the memory is configured into the equipment when manufactured, whereas other data and functions can be programmed on installation to include details of authorized persons and credentials held. Times of transactions and timers for the access functions are also programmed at the commissioning stage. This information must be displayed to the installer to confirm the data that the system is holding and this can be done in a number of different ways. Equally there must exist a method of inputting the data into the memory which tends to be effected by the use of a keyboard that is plugged into the port or physical connection point of the controller or this may be performed by a purpose designed keypad.

The display of the specific information presented to the system will be outputted to either a VDU screen, a printer or a display with LEDs referenced to numbers and letters or to an LCD display. These will depend on the controller type and the system reader. This data function, including the information which is inputted into the controller by the installer or which is outputted by the controller to a display for confirmation is called the data input or output device. It is not to be confused with the general input and output terminals that provide for logic and can switch to indicate events and are used to interface other systems to the access control system.

Observations can show that the reader and controller in Figure 4.3 are integrated in one unit and only the power supply is held in a different location. In Figure 4.4 the reader and controller are separate units as is the power supply, so this makes for a more secure system.

Figure 4.5 identifies a swipe card type reader with a separate power supply and control electronics/switch control unit held in a secure enclosure together with a standby battery. This affords a good degree of security to the control equipment as it is held in an area that is normally not accessible.

This version uses eleven cores to link the reader to the controller, which has three relay outputs.

The wiring and power supply units in all these examples are configured differently and they all have unique characteristics of programming depending on the software they feature. The options available grow in relation to the software features offered by the controller.

The software forming a part of the installation holds the data that relates to the electronic access control system to validate the person

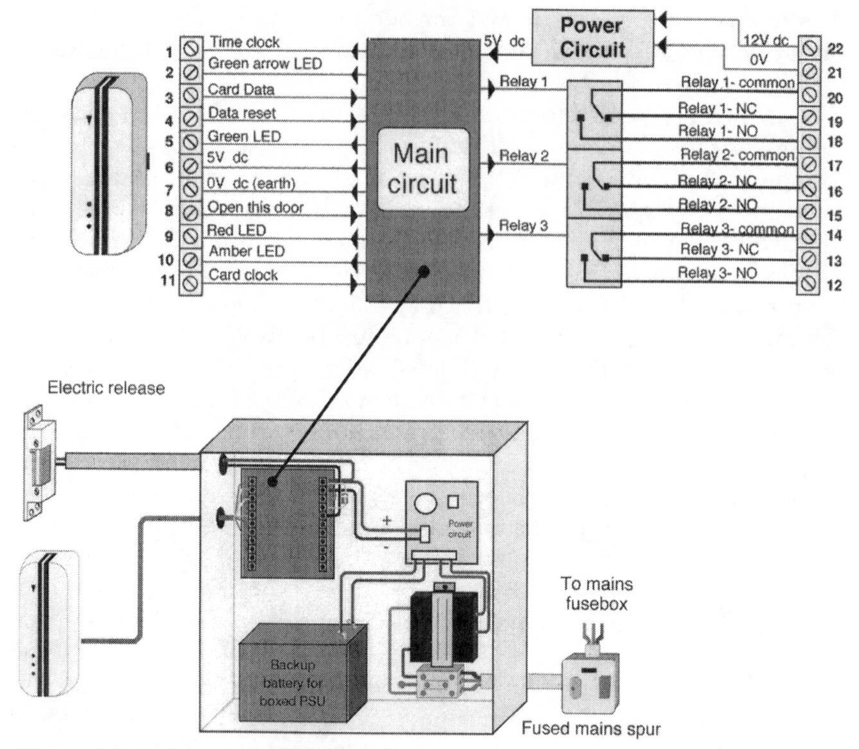

Figure 4.5 *Swipe card wiring technique*

wishing to gain access and is also responsible for the reporting of all the details of the system. The central control or processing unit can be seen as a small computer with memory and comparison programs plus any additional features offered by any given software. The actual complexity is a function of the size of the system, the level of security and data that it must retain. The different systems will be configured in different ways and in a small system it may be possible to find the controller, processor programmer and printer/display unit as a single entity and not as separate components. There will always be a capacity to which the system must work but the following can be used as an example as to the data that should be held as a minimum:

- Name.
- Times access is authorized or time-based entry. This is achieved through a time code or number that represents times of entry, i.e. time code 1 = start 08.30 finish 17.30; time code 2 = start 0900 finish 18.30.
- Card, token, tag, PIN or biometric trait.

- Number of readings before the credential becomes invalid.
- Internal administration reference number. This may be printed on a card or such to help identification.
- Days of the week access is allowed.
- Holiday periods and public holidays.
- Overtime changes.

In the single door system we can therefore say that the controller or processor controls the devices such as readers that then authenticate the transactions, which in turn signal the locking devices or give outputs to other systems and signalling devices. In the larger system and in cases where there are multi-doors to control, a supervisory computer is used to oversee the entire network including any other computers or micro-processor devices that may exist. For the larger building or site it is more difficult to reduce the space requirements and the data that must be held by the software for multi-door systems becomes more detailed.

4.2 Multi-door systems and software

These systems feature a greater number of components than the single door system but all the different door controllers are configured to provide a similar role in that the transmitting of their signals is made to a single position. This means that the format of all the doors being controlled within the network is held at a central point. There are a number of techniques associated with these systems.

- Distributed intelligence. A series of small systems each processing small amounts of data which is then reported to a central point. These small amounts of data may be processed individually at the door controllers to give a measured amount of local control and this is then transmitted to the central point. In the event of a failure of a part of the system it follows that other parts may still function as normal.
- Centralized intelligence. The system functions are all processed at the central point so although the controllers are sited at the various locations they are not locally programmed. There is no duplicating of information with this technique but they can be affected by a failure of the central control point processing equipment. These systems benefit from an ease of programming and there is no measurable duplication of resources.

Off-line is the term used if the central control unit is unable to communicate with the local controllers and this could be attributed to a number of reasons including faults in the local controller itself or in the central unit. Faults can also occur because of damage to the interconnecting cables. For this reason distributed systems should have an ability to unlock doors even if they are off-line and it is of benefit if all controllers can also store information related to the credentials. Certain systems have site codes held within them in order that temporary access can be effected until the original network is restored and this only lowers the level of security in that individual transactions are not recorded. It may also be required that all events are recorded until the network is repaired and it may be possible to achieve this by the use of a memory buffer that can be connected into the system until the central control is once again functioning.

There are many multi-door systems available and these differ enormously in the features they offer; and they also govern the cabling, which is considered in Section 4.5.

In the multi-door reading access control system because the readers used are all remote from the controller and the lock control circuits, a high degree of security is assured as tampering with the reader connections will not affect the door locks. The controllers vary in type and options but can as an example come complete with an onboard keypad for Yes/No programming. This will be equipped with LED and audio feedback so minimum familiarity is needed to set up the system. This controller with its integral keypad can be sited in a secure position so that it is not visible under normal circumstances. Adding or removing credentials such as cards can be done singly or on a block basis using the keypad. Cards or other credentials can be controlled by Learn/Forget cards functions. It may also be possible to unlock doors from the keypad or via the push-to-exit buttons for momentary operation. The mains and low power status can also be indicated on the controller.

Although with many systems multiple doors can be configured back to the controller, reference can be made to Figure 4.6 for an overview of a different technique – the building block principle. In this system the intelligent controllers can each control two doors but are themselves linked together using a two-wire bus. In this specific version up to 63 units may be installed on the network allowing control over 126 doors. This is illustrated in Figure 4.6.

In this case the input is by a hand-held keypad for programming. This keypad can also view system status, alarms and the complete network data. High reliability is afforded as each controller stores the full system database of the network which is transferred to any other controller and in the event of memory failure the unit is automatically refreshed from other controllers.

The network, using two-wire RS485 data cable, can be up to one mile and remote sites can be serviced using a telephone link. This system can also be interfaced by an RS232C serial remote computer/printer port to

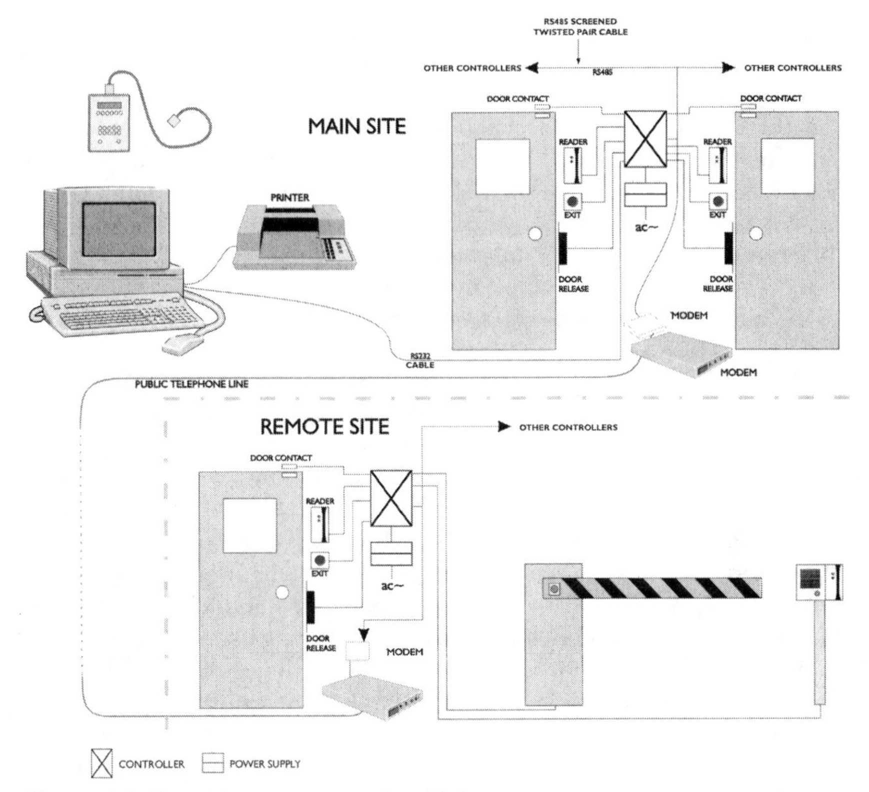

Figure 4.6 *Network access control multi-door.*

allow for DOS or Windows software to be added for total control with the advantage of name files, card holder and alarm transaction storage for audit purposes.

4.3 Central processing and personal computers

Security professionals must be able to have an understanding of computer networks and software because of its ongoing integration with electronic access control.

In the larger access control system a supervisory computer or personal computer (PC) with a monitor and keypad can be used to manage the network. In practice it both issues information to other microprocessors in the system and it receives reports from them in order to store data. The computer therefore oversees the entire system network and can be used to view all the control devices including the locking and any ancillary equipment that may exist for such operations as CCTV monitoring. Since the commands are initiated by a single computer the backing up of data onto a floppy disk drive is important. The floppy disk drives of the computer are designed to accept disks that are inserted into them and are used to store the back-up copies and are removable. The hard disks are not normally removed, being permanently fixed into the mainframe and are where the computer core programs are held. There are in fact many different operating systems in current use and the particular commands to perform the functions will differ between them.

The installer must remember that technology moves at a rapid pace and that those who do not appreciate this are left behind. The vast majority of businesses do use computers so there must be no concern in integrating electronic access with computer networks. Insofar as the engineer is concerned, who has the prime task of being involved with access control and security systems, the main question to be asked is what benefits can the PC bring to the system.

The PC-based access control system now allows the control of literally hundreds of doors to be achieved from different sites and throughout the world. Computer networks allow a great deal of information to be ascribed to systems and not just basic levels of data; this information can be readily amended by uploading and downloading from a remote point. This all comes at a time when access cards such as the smart card are becoming more sophisticated and are being pioneered to include more personal details such as National Insurance numbers and biometric traits, particularly fingerprints.

Customer demand will be the main factor that will drive progression of PC-based systems. This will be to satisfy the needs of the client, and the system and functions that are wanted. Indeed any company with a PC can have this integrated with their access control as the technology of access systems is engineered in line with computer techniques.

It will be understood that access systems are intended to provide a secure means of access but when PC-based, business management can also be incorporated. To this end there are enormous benefits.

Management information improves safety and security and aids business efficiency. It allows the monitoring of access times with customized computer printouts and audit trails of individuals throughout the working day. The exclusion of persons to particular areas is easily accomplished and temporary access for visitors can be readily issued. In essence the zoning of areas and times for access can be changed at random by an authorized management operator.

Total site control with advanced levels of personnel management can go hand in hand with time and attendance monitoring; it can also ensure that the staff are paid accordingly for the hours they work. Using computer networks integration can also play a large part by bringing other technologies in line such as intruder, CCTV, fire and building management. As a prime example, CCTV can keep a visual record of all doors as they are opened and fire alarm systems can be interfaced so that all doors automatically unlock in the event of a fire signal. Reporting can hold a printout of all the persons that may be in the building at any given point in time.

Initial expense can be a problem but this must be balanced against cost effectiveness. There is also more complex cabling associated with a PC network. If modems are installed the more complex computer-based problem can be diagnosed by a specialist from a remote location so the software status can be viewed more quickly in the event of a complication. Practical configurations and experience in the sector can be important. The central unit must always be in a secure area to prevent any tampering and it is advisable to use an uninterruptible power supply (UPS) to ensure that a mains failure problem cannot close down the system, as covered in Section 4.4.

We can therefore overview the supervisory computer as a PC with a monitor and keyboard that is used to manage the electronic access control network as an item that issues information and instructions to other devices in the network and receives reports from them. It then stores information about ongoing events but it is important to make a daily back-up of all files so that a virus cannot cause information to be lost. The computer's bus or circuit board is responsible for the channelling of electrical pulses and signals to their destinations and should be formatted so that it has the most possible available paths to ensure that data transmission is both fast and smooth. The important characteristics of the electronic memory and microprocessors are closely related since they transfer information electronically based on transistor technology. Important is the length of time that usable memory can be retained, released or acquired with the acronyms as:

- RAM or DRAM (Dynamic Random Access Memory). This can be accessed at random and is not controlled linearly such as on a tape.
- ROM (Read Only Memory). This cannot be altered once recorded. A CD-ROM is a compact disc (CD) that holds data which, once recorded, may not be altered.
- PROM (Programmable Read Only Memory). After encoding this is as a ROM but it is a separate chip containing microprocessor information.
- EPROM (Erasable Programmable Read Only Memory). This operates as a PROM but can be changed and programmed in a new format. It is generally found in new microprocessor developments.

In order to make the connection between the control network and supervisory PC, serial communication ports are used.

Serial ports on PCs and workstations in the main use the RS232 standard. It may be noted that this form of connection appears on the equipment in Figure 4.6. The RS232 standard defines the electrical and mechanical characteristics of the connection but successful communication can only be achieved by the software of the devices which are being connected to each other.

RS485 is a further standard for serial connections and this appears on the same figure as a means of connecting the controllers on the data bus.

Ethernet is a system that also allows connection to be made to PCs and workstations by RJ45 ports.

In conclusion, any access control system can be integrated with a computer network to increase the system's flexibility, retention of information and to involve business management. It is important to realize that the access control engineer need only understand basic computer concepts and to recognize that standard computer networks can be modified or bespoke software can be added to the network to form an expert system. When graphics are included to show such features as floor plans, the events of the system or alarm signals can be readily observed so that the maximum amount of detail is available for the VDU operator.

4.4 Power supplies

These may be an integral part of the controller or central control unit or they may be found remotely located. If the system also features remote signalling the power supply should be approved by British Telecom (BT) if it is connected to the telephone network and holds the telephone communication link.

The term power supply can mean a transformer, a battery or a rectifier filter with or without a charging circuit that converts alternating current (AC) to direct current (DC) – we always tend to apply the term to the components as a group. Certain power supplies will have rechargeable batteries included to act as a secondary and standby source.

The operation of the power supply is initiated at the step-down transformer that converts its 240 VAC mains supply to a lower AC source generally in the order of 12–24VAC. The transformer is a device employing electromagnetic induction to transfer electrical energy from one circuit to another, without direct connection between them. In its simplest form the transformer consists of separate primary and secondary windings on a common core of a ferromagnetic material such as iron. When AC flows through the primary the resulting magnetic flux in the core induces an alternating voltage across the secondary; the induced voltage causing a current to flow in an external circuit. In the case of a step-down transformer the secondary side will have a lesser number of windings. From this transformer, power is provided through a two-conductor cable to a rectifier and filter circuit where AC is converted to DC. A charging circuit will be held within the power supply so that a standby battery if specified will be constantly charged so long as AC is present.

The power supply must be voltage regulated and capable of holding a fixed voltage over a range of loads and charging currents. Microprocessor components, especially integrated circuits, are designed to operate at specific voltages and are not tolerant to fluctuations. Low voltages cause components to attempt to draw excess power, further lowering their tolerance, whilst higher voltages can destroy them. For these reasons the voltage should be measured at source and once again at the input terminals on the equipment point.

In access control systems the voltages will typically be either 12 or 24 VDC and the majority of ancillaries work at these voltages including electric locks and readers or scanning equipment.

The critical factor in selecting a power supply is in determining the load it must support. The output must be the same magnitude as a no-load condition to overcome any losses due to cabling and system corrosion resistance. It must be capable of meeting the current demands of all the individual units that together form the system load. The first step is to establish how much power will be required by all of the power consuming devices connected to the supply. If a standby battery is also included the time it can support the system must be established.

In essence we can say that the primary supply is the single-phase 220/240 VAC electricity supply taken to the power supply and used to support the system. The secondary supply would be the batteries if specified in the system.

In practice many access control systems will be detailed with the number of devices that a given power supply can drive and any higher power devices generally come complete with an additional power supply to satisfy their function. However, at times when extra ancillary devices are added to an access system it is vital that the network is not overloaded and that a margin of operating tolerance is added. Remote power supplies can also be used to boost the branch voltages on networks when long cable distances are encountered.

The systems in which we are interested will tend to be powered by a transformer/rectified mains supply perhaps with standby batteries or an uninterruptible power supply (UPS).

The access system therefore relies heavily on the mains supply and it must be of a source that:

- Will not be readily disconnected.
- Is not isolated at any particular periods of time.
- Is from an unswitched fused spur.
- Is free from voltage spikes or current surges.
- Is supplied direct to its intended point and not via a switch or plug and socket that can fail or be switched off either deliberately or innocently

The transformer must be sited in an enclosed position, be ventilated and not fixed to a flammable surface. Transformers tend to be found in the controller or central control unit along with the rectifier and charger unit.

The UPS has a greater ability to negate interference and surges on the mains supply and is used if the system has PC-based networks with back-up. UPSs are used in computer and Local Area Networks (LAN) to protect against irregularities in the input AC supply such as blackouts, brownouts, voltage fluctuations, power surges, spikes and line noise. It effectively interfaces a battery between the supply and the equipment and protects the database if the mains power supply is lost, so it provides much greater protection than standard surge protection devices or power line arresting components. The UPS provides conditioned power to its load and in addition to having far greater protection to interference has increased recording and monitoring. It also has a safety isolating transformer with the specified output plus recharge requirements under any combination of rated supply voltage and supply frequency over the general limits of –10 and 40°C. The UPS will additionally have a low heat output fully rectified transformer, solid state voltage regulator, linear current regulator and high temperature cutout with continuous monitoring of the low voltage circuits. Mains suppression filters are used to combat transient high voltage spikes. The integrated batteries may be sealed lead–acid or Nicad.

In practice the UPS unit should be:

- Of sufficient capacity and recharge rate to cope with any prolonged mains isolation of the mains supply related to work being done for fire safety, normal isolation or normal work on the electrical services.
- Be located where maintenance can be easily performed.
- Have sufficient ventilation afforded to stop gas build-up on the vented battery occurring and causing damage or injury.
- Not exposed to corrosive conditions and that the cells be fully restrained to stop them falling or spilling.
- Marked with the date of installation.

The power supply will be connected to the mains source using cable of not less than 1 mm square cross-sectional area and be correctly earthed. The cables connecting all the other components on the network will be found to be of a number of different types.

Indications on the power supply or controller, if they are integrated, should show mains healthy and faults such as mains failure and/or overload. In physical size the enclosure may be large enough to hold a 12 V 7 Ah battery. In many instances manufacturers quote a period that a particular standby battery can support the system in the event of mains failure, although this may not include the supporting of electric locks as they may need be driven from a different source.

The power supply is a vital component and must be of reliable quality and of adequate capacity to run all of the auxiliary equipment plus the system controller. When selecting the power supply it is necessary to consider the combined surge current of locks and other equipment in addition to the continuous running current as the power supply must be able to cope with an initial power-up and after being disconnected from the mains. It is also recommended that a unit is selected that has inbuilt short circuit, thermal and overload protection as standard as these features can save damaging expensive control equipment. The power supply should also be able to cope with charging the battery without compromising its current rating. Certain power supplies are selectable as either 12 or 24 VDC operation so there is no need to stock different items to drive the various locking devices.

Table 4.3 gives general data on typical sealed lead–acid batteries used within power supplies as secondary supplies or the standby source. It is to be noted that it is possible to obtain higher capacity batteries than 7 Ah for the larger more specialized application. Two 12 volt batteries in series can be used to support 24 volt systems and in the same sense 6 volt batteries can be used in series on small 12 volt support systems. The sealed lead–acid battery should be assumed to have a five-year float life.

Batteries are rated in terms of their voltage and ampere hour capacity. The open circuit voltage of any fully charged lead–acid cell is marginally greater than 2.1 volts so a six cell battery will be 12.6 V. The parameters are:

- Ampere hour (Ah). The product of amperes by time in hours. Expresses the battery capacity.
- Capacity (C). The discharge capacity at a given rate and temperature. The available capacity refers to ampere hours that can be discharged from a battery based on its state of charge, rate of discharge, ambient temperature and specified cut-off voltage.
- Float service. Method in which the battery and load are in parallel to a float charger or rectifier so that the constant voltage is applied to the

Table 4.3 *SLA batteries' typical data*

Weight (kg)	*Nominal capacity* (Ah)		*Dimensions* (mm)		
			L	*W*	*H*
0.31	1.2	6 V	97	25	55
0.57	2.8	6 V	134	34	64
0.87	4	6 V	70	478	105
1.32	7	6 V	151	34	97
1.98	10	6 V	151	50	97
0.58	1.2	12 V	97	48	54
0.7	2	12 V	150	20	89
0.83	2.1	12 V	178	34	64
0.95	2.3	12 V	178	34	64
1.12	2.8	12 V	134	67	64
1.2	3.2	12 V	134	67	64
1.75	4	12 V	90	70	106
2.65	7	12 V	151	65	97

battery to keep it fully charged and to supply power to the load without interruption or load variation.

- Cut-off voltage. The final voltage of a battery at the end of discharge or charge.
- Impedance. Resistive value of a battery to an AC current expressed in ohms (Ω).
- Nominal capacity. The nominal value of rated voltage/capacity.
- Open circuit (OC). Voltage when isolated from the load.

4.5 System cable types

Having earlier looked at some examples of different controllers and systems and their associated wiring techniques it becomes apparent that there are diverse methods of cabling used in electronic access control systems and these differ further when we investigate the competing manufacturers' products.

In the intruder, fire protection and CCTV industries the cabling is more clearly defined but because of the diverse nature and range of access control systems that exist, to include complex networks, a more broad base of cables is used. In addition many specifications now ask for low smoke and fume sheathed data cables to be used as these are safer in the event of a fire. These have halogen-free insulation and sheaths that produce low smoke and fume emissions when burnt. Therefore the wiring will vary between systems depending on the function it must perform and the components with which it has to interconnect.

In many cases the type of cable will be that recommended by the manufacturer to link their various parts based on knowledge of its performance in the field. This knowledge will have been gained historically and may well follow a learning curve yet must still be installed to meet given criteria.

Electronic access control systems must transmit data that can be affected by electrical interference, be troubled by voltage drop over extended distances and be corrupted by capacitive and inductive effects. All these factors govern the selection of cable. Certain considerations should be understood as these apply across the range of access systems.

Communication cables carrying data should be screened to metal parts as this protects the integrity of the information being conveyed to the electronics from radiated emissions in the local atmosphere. It also increases performance over distance and provides the safe passage of static electricity applied at the reader of the system by human contact to the safe earthing point. If push-to-exit buttons and lock release bodies are also of metal these could act as points of discharge and should therefore also be earthed.

Capacitance becomes a principal consideration in longer higher speed data communication in network systems and it causes distortion and error in the communications. It is further aggravated by cumulative cable distance, as this increases capacitance, or by increases in frequency or data speed. The option is to select communication cables with the lowest possible capacitance quoted in pF metre.

For communications transmission the cables and individual conductors can be wrapped with a variety of materials. Such wrappings and jackets include insulation and shielding and strengthening that protects the assembly against stretching during installation. The materials combat dampness, protect from cold and heat sources and provide shielding from electromagnetic interference. In communications the conductor cables contain copper wires and need a ground wire or grounding material within the jacket in addition to the wires that carry the current. The exception to this is fibre optics which do not need a ground.

Power cables are subject to special attention since as the current and distance increase a voltage drop must occur.

The voltage applied to the cable and the voltage at the load differ, therefore a consideration is cross-sectional area (CSA) since as with a larger CSA the resistance is reduced. As an example a cable of CSA

$0.2\,\text{mm}^2$ of R 0.01 Ω/metre is to supply a lock rated at 1 A over a distance of 100 metres. The supply is 12 V.

Total cable resistance = 0.01 Ω therefore over 100 metres $R = 1\ \Omega$.
$V_{\text{drop}} = IR = 1 \times 1 = 1$ volt.
Terminal V at lock must then be 11 V.

If the lock has an operating tolerance of 10 per cent then the terminal V is satisfactory. However, if the lock draws more continuous current or has an increased peak current then the cable length must be reduced or the CSA increased. For this reason installers often double up on cores although long cable runs must always be avoided.

While we can look at the most used cable types adopted for the various functions it must be understood that these are available in a number of gauges of conductor size and with different materials used for the insulation. These depend on the role they must fulfil and the environment in which they must work. Cables for duty outside are subject to special requirements for the outer jacket.

Cable guides tend to class cables in sections for roles applicable to electrical installation, flexible mains, equipment, data transmission, audio visual and coaxial screened duties.

Electrical installation cables are diverse and extend from conduit cables through to fire resistant cables, mineral insulated cables, high temperature industrial cables, signal/alarm and telephone types and include a vast range of screened and non-screened multicores and ribbon cables. Many of these will be used in access control systems and indeed simple door entry techniques use alarm/signal and telephone wiring for some duties. Surface wiring cables for connection to the mains supply also fall into this category

Flexible mains cables meet particular criteria for their insulation properties and will be used for the connection to the power source. Included in this group are some screened cables that can be used for access control duties.

Equipment cables are essentially single wires that may be solid core or stranded and as the name suggests are used within equipment and are not intended for installation practices.

Data transmission cables are not regarded as power cables or cables for the direct connection of equipment to mains power supplies but are designed to meet the special requirements of high speed data transmission systems associated with data processing. Included in this category are those systems satisfying process and control applications.

Audio visual cables have particularly good signal carrying and screening properties and are often also found in computer and data transmission applications.

Coaxial cables are primarily for video transmission and camera control but also extend into access control applications when integrated with monitoring by CCTV.

It is possible to summarize the most encountered cables used in the installation of electronic access control systems.

Unshielded twisted pair (UTP)

A principal form used in electronic access control systems. The cable holds a number of pairs of twisted wire. Each pair is twisted differently and is defined by the number of twists over a given length. The wires are not run parallel with each other inside the jacket but twist across each other as this helps to cancel noise or electrical signal interference that may be induced by adjacent cabling feeding different apparatus. It is a cable type that is in common use in communications wiring and high rate data transmission applications.

Shielded twisted pair (STP)

Also regularly used in communications wiring it is as UTP, holding a number of pairs of twisted wires but these are wrapped together inside a braided copper mesh and enclosed within the outer jacket. It is especially advocated when computer networks are involved. A copper drain wire can run alongside the screen and be used for the termination.

Parallel wire (no jacket)

Twin wire similar to bell wire it has no outer jacket over the insulated conductors. They may be called twin loudspeaker wires and can be used in some ancillary duties.

Twisted pair (no jacket)

As unshielded twisted pair but has no outer jacket over the insulated conductors. These lack the protection of cables with jackets but the twisted construction ensures minimum radiation and reduction of crosstalk between pairs.

Multiple cables with jacket

These form a huge subject and are available in many forms including multiple unshielded wires, multiple twisted pairs, individually shielded pairs and non-twisted wire, all within a shielded or non-shielded jacket.

Coaxial cable

This type has a single copper wire, with the second conductor being the shield surrounding the core. The core is surrounded by a plastic insulation that is then wrapped in the shield of braided copper mesh and then contained within the outer jacket. In real terms coaxial refers to any cable in which the conductor and shield share a common axis. The distance and cable routing of these cables requires careful planning and high quality cables must always be used.

It is possible to obtain coaxial cables with a solid or a stranded inner conductor and the latter is better employed if greater physical flexibility is needed as it bends more readily.

Fibre optic cables

These are a non-conductive cable form that will be increasingly found in communication wiring as they are totally free of electrical interference and can transmit data over extended distances without a need for signal boosters. They are also highly tamper resistant as their transmission is in the form of a modulated beam of light down a medium of transparent glass or plastic fibre. They are used in conjunction with a transmitter that converts an electrical signal to a modulated beam of light, which is then carried by the fibre optic cable to a receiver that reconstitutes the light beam to an electrical signal. They consist of the core of glass or plastic fibre surrounded by a cladding to keep the light within the core and a coating to give protection to the assembly (Figure 4.7).

Fibre optic links are therefore a separate identity to copper conductor cables but can be linked by hubs to traditional cable networks.

It cannot be overstated that there does exist a vast range of cables that can be employed for the different access control duties, systems and the connection of the various units.

Manufacturers' specifications will recommend cables to interface the various components of the system and may also quote a maximum distance for the cable run.

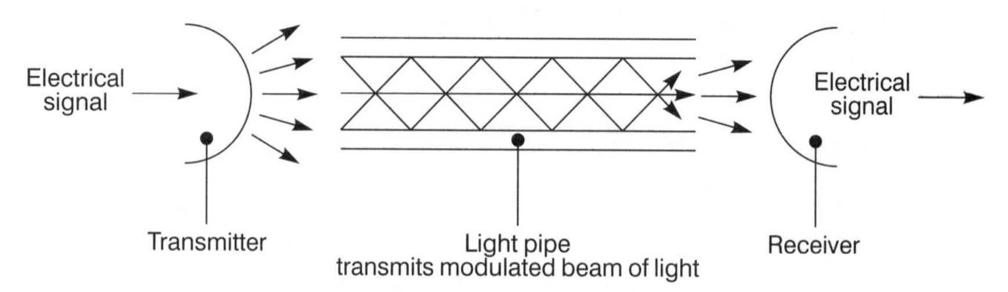

Figure 4.7 *Fibre optic link technique*

Typical cables used for communications, networking, reader connections and PC interfaces are Belden. These are tabled as follows and can be cross-referenced to other manufacturers' exact electrical equivalents.

- Belden No. 8723. Terminal cable. Two pair 22 awg polypropylene insulated. Each pair insulated with an aluminium–polyester shield. Complete with a drain wire. Pairs cabled on a common axis to reduce diameter. PVC jacket.
- Belden No. 9207. Twinaxial cable. Twisted pair 20 awg. Centre conductors each insulated with polyethylene. Overall tinned copper wire braid providing 86% screen with Duofoil shield. PVC jacket.
- Individually shielded pair. 22 awg stranded tinned copper conductors with polyethylene insulation. Each twisted pair is shielded with a Beldfoil aluminium–polyester shield and drain wire. PVC jacket. Nominal capacitance between conductors 98 pF/m.

Belden No.	*No. of pairs*	*Belden No.*	*No. of pairs*
8777	3	8775	11
8778	6	8776	15
8774	9	8769	19

- Multi-pair with overall shield. 24 awg stranded tinned copper conductors. PVC insulated and laid up in twisted pairs with an overall Beldfoil screen for 100% screening with a drain wire. PVC jacket.

Belden No.	*No. of pairs*	*Belden No.*	*No. of pairs*
9501	1	9507	7
9502	2	9508	8
9503	3	9509	9
9504	4	9510	10
9505	5	9515	15
9506	6	9519	19

- Multicore overall shield. 24 awg stranded tinned copper conductors. Insulated with 0.25 mm PVC, overall Beldfoil screen with a drain wire. PVC jacket.

Belden No.	*No. of cores*	*Belden No.*	*No. of cores*
9533	3	9538	8
9534	4	9539	9
9535	5	9540	10
9536	6	9541	15
9537	7		

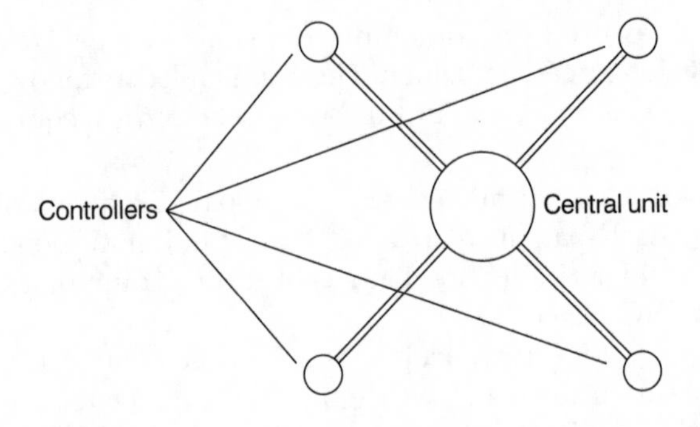

(a) Star. Controllers spurred

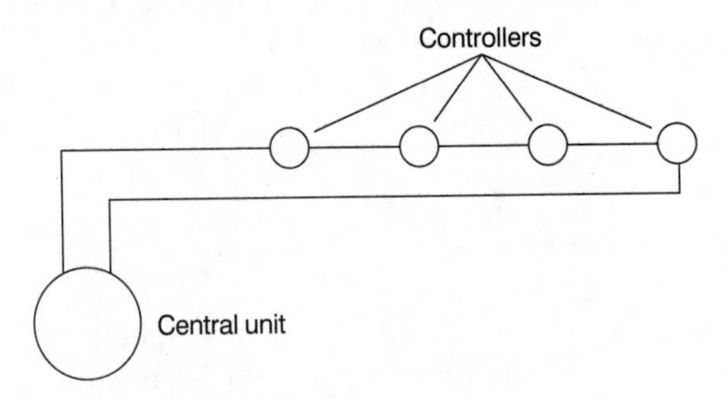

(b) Daisychain. Loop method

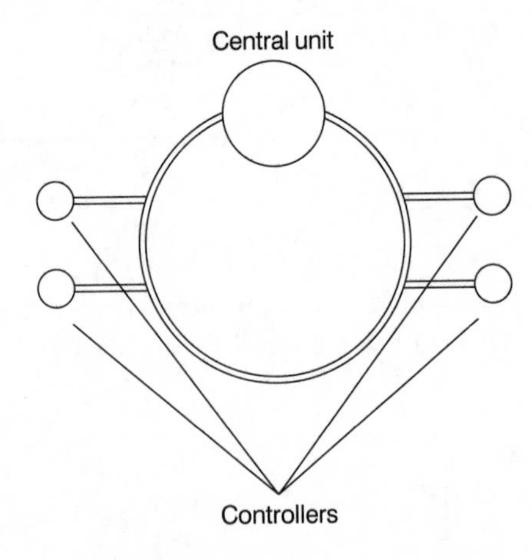

(c) Bi-directional loop. True data highway – ring system

Figure 4.8 *Network cabling techniques*

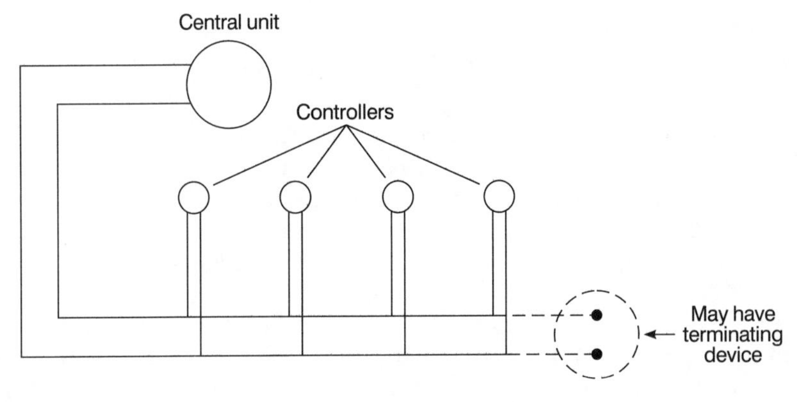

(d) Common bus. Multi-drop. True data highway

Figure 4.8(d)

In addition to a variety of different cable types being used in the electronic access control industry there are also a number of different methods of linking the controllers with the central unit and interfacing all the other devices on the network. The data network can be wired using various techniques that in fact use terms that are also associated with other electrical industries. In the same way as recommending particular cables different manufacturers will also advocate the preferred technique of cabling the controllers to the network. These controllers may be called nodes, which is a generic term for a device connected in a data network and is used to great effect in the intruder industry. In practice the node will have a transmitter and receiver contained within it and be connected by a star network, daisychain, bi-directional loop or common bus multi-drop method. Certain systems can be cabled using a mix of the various techniques (Figure 4.8).

Star network

This is easily understood as it is effectively a technique by which all controllers or nodes are spurred off from the central control unit. In practice all of the cabling is routed direct to the central control unit from the controllers so it does not necessarily qualify as a true communications network because a common data highway is not used. It involves a greater deal of cabling if the central control unit is not geographically central to the nodes, but the cabling is inherently reliable as the individual devices are cabled independent of each other. It is easy to add or remove controllers using a star network and a lesser degree of communications intelligence is required in the nodes.

Daisychain

Often also called a loop method it is cabled with the transmitter of the first node connected to the receiver of the second node. The transmitter of the second node is itself cabled to the receiver of the third node and so forth until the transmitter of the final node in the network is connected back to the receiver of the first node. Using a daisychain method any transmitter can correspond at any given time. This is possible as its data is picked up by only one receiver, although the data is given usually in one direction only. The daisychain technique allows for long distance transmission because signal boosters are incorporated in each node, but a failure of one part can cause problems throughout the entire network. In a similar way to the wiring of the star network this method does not make the most efficient use of the communications bandwidth.

Bi-directional loop

This is similar to the common bus technique using a data highway but all of the controllers are spurred off a loop that returns in a similar fashion to a ring main. It makes maximum use of the communications bandwidth, but high intelligence is needed of the central control although lesser levels are required in the nodes.

Common bus

Also called multi-drop this technique is popular because it is not difficult to add or remove a node. It uses a true data highway so that a failure in any controller or node does not affect the remainder of the system. In this method there is a high level of communications involved making maximum use of the bandwidth but there are limitations on cable length. It is cabled using one wiring run so that the transmitter and receiver of each node are connected over the common cable together with the transmitter and receiver of all other nodes in the network. There may be found a terminating device at the end of the cable. In larger networks repeaters may be required to boost the signal. The common bus makes maximum use of its cabling and saves on installation times. A major consideration is the integrity of the cable – if this is destroyed or damaged the system is lost from that point to the end of the cable run.

Therefore, consideration must be given as to the system integrity if a node fails or a cable is destroyed or damaged. To this end the use of either distributed intelligence or centralized intelligence systems must be balanced against the risk, probability and consequence of failure.

It follows that the cabling is a vital part of the system so the protection of it and its filtering should form a separate subject.

4.6 Cable protection and filtering

There are precautions that can be taken to minimize any damage to cables being made as an addition to cables that have integrated protection such as armoured jackets. There is a need also to protect the cable against susceptibility to different sources of induced interference.

Regulations cover the segregation of cables of different categories and the susceptibility to voltage surges, radio frequency interference (RFI) and electromagnetic interference (EMI), but good practice must equally be applied to ensure the neatness of the finished cabling system and the mechanical protection of the vulnerable wiring. We can briefly overview the techniques.

Cable protection techniques

At this point we only wish to confirm the most popular forms of cable protection techniques or containments. Reference should be made to Section 6.1 as this includes information on the installation of the cabling plus data on fixing and spacing.

Steel conduit

This is available in standard sizes of 20 mm or 25 mm diameter. It has great strength, is waterproof, is mechanically continuous, provides earth continuity and is non-flammable. The disadvantages of steel conduit systems are that they can form condensation within themselves and are subject to corrosion.

Non metallic conduit (PVC)

Also available in standard sizes of 20 mm and 25 mm, these systems have high resistance to corrosion by water, acids, alkalis and oxidizing agents or by the chemical components in concrete and plaster. They are dimensionally stable, non-ageing, non-flammable and have a high electrical breakdown voltage but have a low resistance to cutting.

Trunking (PVC)

Standard sizes in mm are 16 × 12.5, 16 × 16, 20 × 10, 25 × 12.5, 25 × 16, 38 × 16, 38 × 25, 38 × 38. Trunking is at an advantage in that unlike conduit it can be applied to existing cable systems without a need to disconnect them. Its other advantages and disadvantages are similar to those of PVC conduit.

Although it is possible to obtain steel trunking it is rarely used for access control signal and communications wiring but tends to be used for protecting mains cables in industrial areas.

Aluminium tubing

The standard size is 12 mm diameter. These systems are neat and have good resistance to corrosion and condensation but they have a low resistance to cutting.

Capping

Capping is used to give a level of protection to cables that are to be contained in the building fabric. They are available in many sizes but are not intended to be used to enclose cables unless they are to be buried at the final stage.

Filtering

The cables that are to be used and the means by which they are connected and terminated have an effect on their resistance to suppress certain forms of electrical interference.

In Section 6.3, which deals with the connection and testing of the system components including the interconnecting cables, we consider how some types of interference should be suppressed by use of screened cable and single point earthing. There are, however, many ways in which the network can be affected and different techniques to try to negate these effects.

Lightning and voltage surges

All mains power lines entering or leaving a building are prime routes for lightning and switching induced surges and overvoltages to enter the electrical and electronic systems. Due to the large cross-sectional areas of these cables high current and voltage surges can enter with minimum attenuation.

By fitting surge protection devices in parallel with the incoming mains, surges and overvoltages can be reduced to an acceptable level. Protection devices can be applied to single- or multi-phase systems although the access control system should be supplied by the same phase even if various components are spurred from different points in the network. The increased sophistication of modern electronic security systems calls for reliable and uninterrupted operation so the effect of a lightning induced surge or overvoltage entering the equipment will cause malfunction or destroy the processing control unit. The surge protection device can also operate and reset automatically so that the need for manual resetting is eliminated.

The effects of lightning and voltage surges are the corrupting or erasing of data and also the damaging of equipment. Lightning is readily understood as a problem area and easily defined. However, voltage surges are essentially the effects of electrostatic discharge, inductive load and power grid switching. Lifts, copiers, fluorescent lights, vending machines, computer printers and such are examples of equipment that can cause transients and slow degradation in equipment performance.

In practice, voltage surges occur many times a day on typical AC power and communication lines so it is vital to take measures to prevent damage as a result of them. At the present time this is most apparent as increasing numbers of companies depend on the highly vulnerable circuits of computers and security systems.

Uninterruptible power supplies (UPSs) cannot protect sensitive equipment against all transients and these themselves may need protection.

There are a number of general practices that can be carried out to overcome the problems attributed to surges:

- Add surge arrestors to the secondary side of transformers.
- Avoid external wiring.
- Connect all steel conduit tubing to an effective ground.
- Provide a separate ground heavy gauge insulated earth to the main system transformer.
- Add unique protectors to telephone socket outlets intended for modems.

BS 6651: 1992. Protection of structures against lightning, addresses the problem in Appendix C. This increases the scope of the standard to include guidance on the protection of the equipment in which we are interested and also gives advice on assessing the level of risk. It recommends methods of protection and the selection of protective devices. The appendix also takes account of the effects a lightning strike can have for buildings with external power and data feeds and how they may be affected by strikes anywhere near these cables. The appendix is concerned with cable runs up to 1000 m from the building.

There are actually three areas in a building with different protection requirements:

Category A	The load sides of socket outlets
Category B	The mains distribution system
Category C	The supply side of the incoming distribution board

Table 4.4 *BS 6651. Magnitudes of surge voltages/currents to be protected against*

	Category		
	A	*B*	*C*
Peak voltage 1.2/50 μs wave form	6 kV	6 kV	20 kV
Peak current 8/20 μs waveform	500 A	3 kA	10 kA

Category A may not always be applicable as the socket outlets may be too close to the distribution board and in such cases a category B protective device should be used.

The magnitudes of surge voltages and currents defined by BS 6651 that should be protected against at different locations on an arc power system and in an area of high system exposure are given in Table 4.4.

System exposure is affected by several factors such as geographical positioning and the level of loss inflicted if a lightning strike was to cause damage.

It is a requirement of BS 6651 that all telecommunication and data communication equipment must withstand those surges defined in category C, i.e. 10 kA peak current for high risk system exposure, but this is reduced to 5 kA for medium risk exposure areas. The reason for this is that the electrical characteristics for power lines and signal wiring are different in that a surge travelling in a signal cable is not attenuated by the cabling in the same way as a power line surge. Clearly it is the surge protector which must withstand these peaks to ensure that the equipment continues to operate, but it must itself survive the transients and resulting currents, allowing through only the residual part of the surge which is not capable of causing any damage.

The residual part of the surge is called the 'let through voltage'. Most electronic equipment will withstand short duration surges of around 1250 V on the mains supply and a reasonable safe margin is a let through voltage at the equipment interface of not more than 900 V. Clearly the closer the let through voltage is to the equipment supply voltage, the better the equipment protection.

Mains protection systems

There is a wide range of transient voltage surge suppression products available ranging from surge protection systems based on a trilevel zoned protection strategy that starts where power and communication lines enter a building to special protection devices for extra sensitive equipment. The heart of these systems is a distribution surge protector (DSP) which is installed at the cable entry to a building and at the distribution point for each floor of the premises. There is also a range of devices available that incorporate power and modem protection in one unit.

We can say that transient suppressors in the form of fused spurs can prevent the distortion of voltage, current and frequency as a result of transients. If these devices are used local to the control equipment in place of the standard spur together with a transient surge protector wired across the building mains supply the problems of mains interference and the resultant damage should be negated.

Data line protection

As the access control engineer becomes increasingly involved with remote signalling and data transmission the protection of data calls for a somewhat different approach to the protection of power lines.

Standard hybrid surge protection devices are usually not designed for high bandwidth data transmission as well as significant surge diversion capability. Also, the need for low clamping voltages and delicate impedance matching makes circuit design particularly difficult. This applies even more when high transmission speeds are specified.

Specialized protectors are, however, produced for telecommunications areas and modems, otherwise these can be severely damaged or even destroyed by large transients. To this end unique protectors are available at telephone outlet sockets intended for modems. The use of other protection devices is advocated where data is to be transmitted via networks between buildings.

In addition to transients on the mains supply there are two other types of interference that can cause problems for the electronic access control system. These are namely RFI and EMI.

Radio frequency interference (RFI) and electromagnetic interference (EMI)

The effect of both RFI and EMI is a distortion of the current flow particularly on DC causing systems to crash or the memory to erase. This is because in the case of equipment such as microprocessors this distortion is seen as a pulse of instruction.

The causes of RFI and EMI differ but are essentially:

- RFI. Induced by transmitters, boosted CB radios and discharge lighting. Low flying aircraft and powerful radar can also cause problems.

- EMI. Generated by radiated electromagnetic energy as mains borne interference. Sources include welding tools, starting and stopping of motors, air conditioning units, any equipment containing a motor and fluorescent lighting.

Although the manufacturers of electronic equipment are responsible for the suppression of their goods at source to combat interference there are additional and certain measures that can be taken by the installer of electronic access control systems. These are:

- Primary filtering of the control equipment using suppressors or stabilizers.
- Screening and shielding by the use of grounded conduit tubes or shielded cable (a).
- Bonding and earthing so that the equipment is shielded by metalwork and the radiated energy is hence conducted to an earthed point.
- Secondary filtering of signal and communication wiring.
- Segregation from other cabling in a building (b).

  (a) Screening and shielding is essentially the enclosure of system parts in conduit and metallic structures that will ground radiated energy travelling through the atmosphere. Screening is achieved by the use of braid in cable, seen as a screen of fine wires grounded to extra earth points.
  (b) Segregation is the use of exclusive routes for data cables to ensure that harmonic currents from other cables, particularly power wiring, cannot be induced into the signal and communications cables. These access control cables are not to be run close to power cables, or in parallel with them but should cross at right angles or be shielded by an earthed division.

As an overview to filtering we can say that this in practice forms a huge and complex subject because of the increased use of data and communications networks plus the increased use of electronic component circuits within all security electronics.

Equally there exists a number of different voltages and frequencies in most buildings so the need for circuit protection and filtering has and will become more of a necessity. The important steps to be taken in all installations therefore must include:

- The screening of the control panel and all of the interconnecting wiring.
- Correct grounding and fusing.
- Filtering as appropriate.
- Protection from lightning strikes.

In the event that long range links between buildings are wanted an option is to consider optical fibres as these avoid the risk of lightning damage and electromagnetic interference or the noise that can be generated in electrical conductors by stray electromagnetic fields.

Optical fibres and their technology have been broached in Section 4.5.

It is prudent at this stage to have an understanding of the Electro Magnetic Compatibility (EMC) Directive because of the way electromagnetic interference can cause equipment malfunction in any sensitive electronic goods. We will also note the use of the CE Mark on many of the products in which we are interested.

EMC Directive and CE Marking

This Directive is relevant to all types of electronic equipment including access control and security products and was introduced to ensure that electronic goods sold within the European Union or other countries do not cause excess electromagnetic interference or are unduly affected by it. In practice every electrical system can act like a radio to some extent since each wire or strip of contact is in effect an aerial that is capable of transmitting or receiving.

The control of emissions has been common practice in military equipment for many years but the huge expansion in the use of electronic equipment for general use has now made the restriction of interference emissions necessary. There is also a need to have control over the susceptibility of electronic equipment to interference from external sources.

Electromagnetic compatibility (EMC) is achieved by the control of the design and manufacture of electronic systems to restrict first their transmission of spurious signals and second their suppression of interference from radio signals. EMC also covers the susceptibility of equipment to spikes in the system power supplies plus static DC charges.

This interference refers to any electromagnetic disturbance or phenomenon which may degrade the performance of a device, control equipment or system. The disturbance can be in the form of electromagnetic noise, a signal or a change in the propagation medium itself that can contribute to equipment malfunction.

The most important item relating to interference within most electronic goods in which we are interested is the microprocessor. The CE Mark (Figure 4.9) is applied by the manufacturer to signify that a product conforms to a particular European regulation. In the case of access control systems the application of the mark denotes conformance with the European EMC Directive. It can be found on the apparatus itself, the

Figure 4.9 *CE Mark*

packaging, installation instructions or on a certificate accompanying the goods.

For any produce carrying the mark the manufacturer is responsible for issuing a declaration of conformity to claim compliance with the Directive. This must be made available to any customer on demand. In addition, the manufacturer of the goods must be able to give evidence to support the declaration of conformity – usually in the form of a test report. In the case of any access control systems products that have transmitter/receiver characteristics such as microwave technology, testing must be performed by an accredited test house. This report must then itself be verified by the Radiocommunications Agency before compliance with the Directive can be claimed. For other products, the manufacturer can self-certify by doing in-house tests at the point of production. In these instances the goods are marked CE0192.

4.7 Discussion points

In this chapter we have found that the system capability of an access control system will vary from a single and simple intercom speech unit with low security to a true full distributed intelligent network controlled by a central PC. The latter system can have extensive alarm transactions, user log and roll call reporting and provide essential management and security information.

The network to be specified will have an influence on the cabling plus its protection and filtering.

The engineer must establish exactly what is required of the proposed system and come to recognize that as the network becomes increasingly complex the input of more persons is required to oversee the installation. These to include not only the representatives of the users, but architects, manufacturers of the equipment and ancillary devices plus safety personnel.

Discussions must cover all aspects and eventualities otherwise it is not possible that the system can be satisfactory as a whole.

5 Survey and design of systems

So far we have looked at the components used to build up a system to include barriers, doors and the control and reading equipment plus the cabling and data networks associated with it. It is now time to consider how information should be collected as a means to determining the way in which the intended access system is to be specified. Following this it is possible to fulfil the survey and produce the specification applicable to the site.

It is necessary to design the system with all the required elements combined before the installation can be considered. Important are the selection of cabling as this has an effect on all of the functions and the consideration that must be given to failure of the mains supply or breakdown of the network and how the system can then operate off-line or with standby supplies. In this chapter we can also turn our attention to sensing devices plus remote signalling that can form part of the design. The engineer must seriously consider integrated systems since the access control techniques may need to be amalgamated with other systems of security, automation or other building services.

The subject of integration is indeed of vital importance as the electronic access control industry is only one part of a much greater security industry. Equally the need for confidentiality in the design and implementation cannot be overemphasized because the engineer must be aware of the risks to themselves, their employers and the customer, so it is important to respect every person's role in the overall scheme.

The reason we see survey and design as amalgamated with integration of systems is because we can see many changes in the security industry and advanced technologies combined with stricter standards and working practices will continue such a trend. Technologies in the security area will become more complementary, and the engineer involved with access control must understand how this will affect him or her in the long term. We can expect the installer's role to develop with verification techniques and CCTV becoming common with a higher profile being put on the external protection of sites. Increased capital is to be spent on making sites more secure and training will advance as customers seek management systems. This will all be a natural progression as buyers seek computer-based systems. The engineer must survey and design with all these thoughts in mind.

Certainly there is an increased demand for perimeter systems and integration will play a major role, although standalone technologies will become outdated, certainly in the commercial and industrial sectors. All this can then surely expand into the domestic scene albeit at a lower level.

We will also witness a growing use of PCs in the installation business and as more computer-based systems come on line the days of traditional programming will become numbered. Software developments will be achieved remotely by central offices using modems with uploading and downloading programmes.

Standards, policies and regulations will change as EN standards are adopted, but as the policies are enhanced it will be necessary to dedicate more time to administrative work. Equipment will become more technical and there must then be a corresponding increase in training particularly with software. Systems will become larger and with the complexity of software and integration the engineer must be alert to the future. However, systems will become refined as the cost of components falls. Architects and planners must now see the future and design for it and see how they can offer additional security measures for prospective access control clients.

Electronic security systems will become a part of building management or intelligent buildings, but the emphasis will come not only from the security industry but from architects, planners and developers. Security installers must react to these changes with regard to enhanced communications and the uploading and downloading of data. All installers must diversify so while we can stress that the security industry is strong we must expect advances in integration and technology with compatibility of system hardware.

5.1 Surveying

In the first instance we should understand that an electronic access control system is the collection of various components that will work in conjunction with each other to control the access to a building or site. These components will also be used in different systems and environments but be planned to work in different ways, therefore the principal task of the survey is to establish:

- The reason for the system.
- Expectations.
- Means of operation of the system.
- Budget.

The answers to these questions will always lead to an overlap based on the aims that are to:

- Provide access to specific areas of a premises or building for those individuals authorized to do so at given periods of time.
- Restrict access to unauthorized persons.
- Tender for a purpose and simple to use and maintain access system.
- Provide the adequate level of security for the premises.

The purpose for installing the system must be clearly identified and the role that the system is to play must be clearly defined. The client must understand the benefits that will be achieved once the system is installed but equally appreciate the demands that must be made to achieve that end. The integrity of system security still relies upon the authorized persons who make use of it and it cannot be regarded to afford the perfect level of security in every detail.

Following the establishing of the role that the access control system is to play it then becomes possible to assess the data needed by gathering the necessary information. Such conditions in a general sense can be:

- The number of barriers or doors to be governed and how these are to be used in practice. It must be known if these are to operate as entrance, exit or as entrance and exit.
- The number of users as this influences the equipment type which has a certain capacity. Consideration must also be given to the special needs of some staff, the turnover rates and the ease of system use.
- The rates of pedestrian or vehicle flow through the controlled points at different times of the day.
- Determining of the area of the building or space to be protected and what materials are used in the construction of the perimeter boundaries.
- The system type as the chosen form must be suitable for the application being considered.
- The physical strength of the barriers or doors and is this adequate.
- If hands free is an option.
- The data that must be gathered such as time of entry, identity of user and control of the times of entry can be granted.
- Level of security to be afforded.
- Provision for removal of credentials.
- If security management software is needed or integration with other techniques is to be effected. Future expansion and upgrading must be allowed for.
- Cabling routes should be judged as these have an influence on installation costs.

- The location of all of the control equipment plus any extra power supplies or remote signalling equipment.
- Operation in the event of mains failure, a partial breakdown or emergency exiting.
- Other considerations involve regulations for fire, emergency operation and the satisfying of any inspectorate bodies.

In reality the access control system must rely on the users for its integrity, but it remains that the surveyor must understand his customer and needs to ensure that the proposed system is correct for its purpose. Indeed the survey forms a huge subject and may be complicated by the budget of the client. For this reason it may be necessary to offer different options but take into account that the client can expect a return from lower levels of theft, reduced insurance premiums and an improved level of staff security.

The systems will vary enormously in use but all of the foregoing conditions should be assessed prior to the design stage. Important is the need to get the correct balance in the system as a whole. Electronic access control engineering is a growth industry and will influence the ways in which all properties will be secured but may need to be supplemented with other technologies if high levels of security are required to include alarm generation and perhaps CCTV monitoring. Perimeter protection is important and barriers offer only limited restriction of access to pedestrian traffic. For door protection the surveyor must look back to basics and recall that rim locks (also called surface mounted locks or nightlatches) are fitted to the surface of a door. Mortice locks are fitted into a chiselled slot, or mortice, within the door's main body and are less vulnerable to being forced. Locks fitted with latch bolts can be operated using a handle or knob and are sprung so that the door can be closed without being locked. Mortice deadlocks can only be operated by use of a key or electrical pulse.

Monitoring of doors is a priority otherwise a system with complex software fails a basic need and the perimeter can be easily rendered worthless. We must reiterate that the system must be considered as a whole and although emphasis should be placed on the main equipment every component still has a role to play.

5.2 System design

Following on from the survey it is then possible to consider the design of the access control system. It is the practice to hold the data on information sheets and drawings so that it can more easily be referred to. The stages of recording the system design criteria can be typically as follows:

- Perimeter and door protection drawings and charts.
- Cabling.
- Security level.
- Mains failure.
- Breakdown.
- Emergency access.
- Number of staff and turnover.
- Software.
- Integration and upgrading.

The first stage should be to finalize a drawing or floor plan of the full site and then to identify on this the individual areas that require protection. The actual access points to be controlled can then readily be determined having received confirmation from the client. The hardware can be recorded together with the barrier and door data in a further chart.

Perimeter and door protection drawings and charts

In order to detail the site and the perimeter protection afforded in drawing and chart form Figures 5.1 and 5.2 can act as an example of a method that can be employed.

Figure 5.1 gives an example of a site layout. In this case the site is a new development and is intended in the main to register staff coming to work and the number of vehicles that will be on site. The vehicles must enter

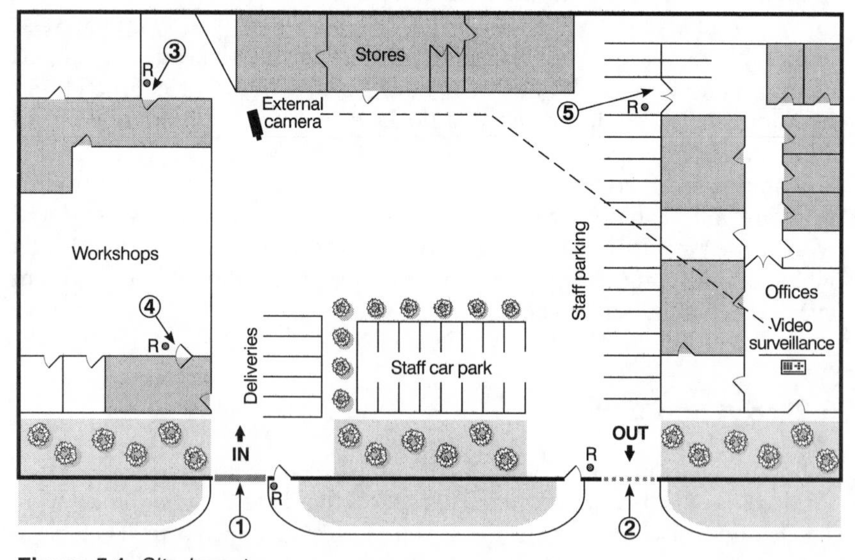

Figure 5.1 *Site layout*

A Barrier/door identification reference number. Internal (int.) or external (ext.)
B Type of barrier/door
C Construction and material of barrier/door
D Frame or barrier support materials
E Barrier /door movement type
F Barrier/door essential dimensions
G Notes

ID. A	Type B	Const. C	Frame D	Movement E	Dimns. F	Notes. Drg ref. G
1 ext.	Industrial rising arm barrier	Steel channel	Steel channel	90° arc	5m span.	See Drg. XYZ
2 ext.	Rising kerb	Steel	High density concrete	0.3 m rise	2 m wide	See Manufacturer's Drg. ABC
3 ext.	Right-hand door	Solid wood	Solid wood	Open in	50 mm thick	Lock at right-hand side
4 int.	Left-hand swing door	Hollow wood	Solid wood	Open in/out	50 mm thick	Lock at left-hand side
5 ext.	Double doors	Aluminium	Aluminium	Open in	75 mm thick	Lock at left-hand side

Figure 5.2 *Perimeter protection reference chart*

by the industrial strength rising arm barrier and exit via the rising kerb. The staff that attend on foot also gain access at the same point and exit through the rising kerbs but in both cases by adjacent gates that are not locked during office hours. The barrier and rising kerbs only restrict vehicle access so their operation is disabled in the event that the traffic is pedestrian only.

Figure 5.2 can be used to show how a chart can be compiled to detail essential materials and other barrier or door data by allocating numbers to the barriers and doors. The chart is to enable the perimeter to be quickly viewed in terms of barriers and doors. Further and specific information can be held on additional drawings.

This example can identify how in the working environment staff can be logged in and out of work by using a card technology access technique. Once inside of the perimeter PINs are used in coded keypads to obtain access to the office and workshop areas outside normal time zones. During normal working hours all doors can be left open.

It is possible to increase the level of security by the use of an external camera and video surveillance with the VCR equipment in a secure point protected by an intruder alarm.

It follows that the coded keypads in other situations could restrict access at any time if the level of security demanded it. Figure 5.3 shows

Survey Company	Client's Address
	Date of Survey
Surveyor's Name	Contact Name Tel. No.
Type of Contract New ☑	Update Existing ☐
Required Completion Date of Installation	
Project Description Log staff in and out of site. 80 staff tokens. Recent access to workshops and offices by PIN s out of office hours. Software to log all events. Install ext. camera & video surveillance and intruder.	
Main Contractor	Drawings Attached ☑
Acceptable Price Range £	
Barrier No.1 In only. All staff to log in. Rising kerb No.2. All staff to log out. Doors Nos 3, 4 & 5 to have manual overrides. New site development. All cables in ducting. Central controller and PC with intruder detection Main Office.	

Figure 5.3 *Survey description*

how a brief survey form can summarize the requirements for the client. Reference can be made to Chapters 1 and 2 for details of barriers and doors in common use.

Once the data for the barriers and doors has been recorded on the information sheet it is possible to supplement this data with the location of the reader or scanner by recording the positions of these as inside or outside of the perimeter protection. It is also advisable to record if they act in an entry or exit mode or in both.

Cabling

It is also important to place on record the cable runs and the proposed routes that they are to follow. In addition the cable types may be detailed as this is an aid to fault finding in the future. Initially it is good practice to look for the routes used by the existing services but to ensure that segregation from other wiring is made. If the cabling is to be made between different buildings it can be achieved by the use of catenaries that go overhead or through underground routes. The integrity of the cabling will always have an effect on the security of the system but this is diminished if cables are exposed and accessible.

A vital part of the cabling system design is the consideration of voltage drops in large installations. Allied to this is the need to determine the standby capacity of power supplies and if there is a requirement for any additional remote power supplies to support ancillary equipment or distant equipment current loads. In Section 6.3 we consider the connection and testing of system components plus the calculation and logging of voltages at distant loads to prove that the voltage drop that will occur from the supply terminals to these loads cannot interfere with the satisfactory operation of the equipment.

It is important that the cables carrying power circuits must not have a quoted resistance that creates a problem for supplying distant loads. Although installers often double up the supply cores to reduce the resistance of the supply conductors this is not always an option. Therefore judgement to distance must be made as the greater the distance or the smaller the cable size, the greater is the resistance, and as the voltage at the terminals of the supply is fixed, voltage must be lost over the cable run.

The head voltage at the load is the terminal voltage less the sum of the voltage drops in the circuit. This should be calculated at the design stage and verified in practice at the testing stage. Manufacturers' data and installation sheets will provide information on the voltage level and current consumption for their equipment or components.

Reference to Table 5.1 illustrates the quoted resistance per 100 m of twin cable and the voltage drop over this distance when a current of 200 mA is

Table 5.1 *Voltage loss at 200 mA*

Wire size	*Quoted resistance/100 m* (Ω)	*Voltage drop at 200 mA* (V)
1/0.8	7.2	1.44
7/0.2	16.4	3.28
13/0.2	8.8	1.76
16/0.2	7.2	1.48
24/0.2	4.8	0.96
32/0.2	3.6	0.72
30/0.25	2.4	0.48
40/0.2	2.9	0.58
0.5 mm^2	7.2	1.48
1.0 mm^2	3.6	0.72
1.5 mm^2	2.4	0.48
2.5 mm^2	1.4	0.28

drawn. The table covers some popular cable sizes and typical quoted resistances.

The current in a circuit can also be predicted when the terminal voltage is known together with the circuit resistance. If, as an example the terminal voltage is 13.8 V and the circuit resistance is 66 Ω then by using the equation $V = IR$ we can calculate that $I = V/R = 13.8/66 = 0.209$ A or 209 mA. Such calculations can then be verified at the testing stage to ensure that there is no fault to earth or circuit leakage.

We can also establish when V = voltage in volts, I = current in amps and R = resistance in ohms (Ω) the voltage drops across cables or the maximum permissible cable resistances.

A further consideration is the total demand current, which is the sum of the current consumptions of all the devices being powered in their most demanding condition. This must be subtracted from the power supply output to ensure that there is an available excess margin.

However, resistive effects can still cause faults at a later stage if corrosion occurs or if cables are badly terminated as this effectively increases the cable resistance and in proportion decreases the voltage over the cable run.

Although this attenuation creates problems for any powered device it can seriously corrupt the supply of data in communications wiring. Resistive effects, if not intermittent, are easily detected but capacitive problems on cabling are more complex.

In data communications voltage pulses are used to convey bits of information and to sense the state of inputs. However, this capacitive effect signal transmission is not easily measured and attenuation of the signal can cause data to be corrupted or to be unreadable. For this reason certain systems call for the use of low capacitance cable for extended distance as these have improved transmission characteristics for high speed and longer distance data transfer. Reference to Section 4.5 identifies common cable types.

Security level

The levels of security for all of the different points within the system may indeed differ and the different levels of risk can be identified on the data sheets.

If interlock accessories are used to make the installation of security door (airlock style) systems, ensuring that one door is closed before another can be released, these devices should be identified on the plans. Provision is made to detail whether they are used with monitored releases or unmonitored releases.

The security can be classed as high, medium or low and can be established with the needs of the client. These levels can be marked on the drawings.

Mains failure

An option is to provide battery back-up for a period selected between 1 and 24 hours longer than the longest unmanned period on the site. The back-up periods can change for the different areas but it is normal to have the central reporting point supported for the same period as the controller. This is due to the controllers reverting to a lower security mode if they are separated from the central point.

It is often the case that VDU screens are allowed to power down to save energy but the central processor is to remain powered. It is also normal that external doors and barriers fail in a locked condition (but have manual overrides) whilst internal doors fail in an unlocked state as determined by the level of security specified for the particular area. The standby supplies must be adequate to support critical components for a prescribed period of time and it is necessary to determine the supply voltages that will be provided by the standby power supplies in the event of mains disconnection. This should be determined in conjunction with the design of the cabling.

The charging voltage and regulated power supplies are essentially higher than those of batteries incorporated within them so this must be taken into account.

Warning buzzers can be installed at remote points to notify personnel of the mains disconnection so that early remedial action can be taken.

Breakdown

This can be considered in the same way as mains failure but can be as a result of cable or component failure.

Centralized systems will have all operations terminated and perimeter devices will remain in their state prior to the breakdown.

Distributed systems will be degraded and off-line but there will remain certain functions that can be performed depending on the system and component sophistication. Prime examples are the granting of access using such features as a site code and the storing of events in a memory buffer.

Emergency access

This is allied to mains failure and breakdown as the failure can occur when the premises are vacant. For the purpose of gaining entry to the site high security manual overrides are used and the location should be specified on the working drawings.

Consideration can also be given to the use of firebreak relays that can switch door release loads and are designed for use on fire exit doors where security is also important. They enable release of locking devices from the fire alarm signal but reset separately from the fire detection system thus avoiding accidental reset during an emergency.

Number of staff and turnover

This has a major influence on the system, for example at peak periods such as start and finish times when large numbers of staff must pass through the readers at the same time and if all events are to be recorded.

The time to process an entry is governed by the time to approach the access point, present the credential, validate the transaction, release the perimeter protection, enter the required area and relock the perimeter protection.

The times to perform this role differ enormously in relation to the identification method. In general the non-contact recognition system and card swipe techniques are the most rapid whilst card insertion techniques and keypads are progressively slower. Personal identification or biometric systems require extended processing periods and are not capable of rapid turnover.

The balance of security and rapid processing must be considered carefully. The staff element related to special needs must also be appreciated. Personnel may be in wheelchairs, partially sighted or infirm. Staff may also need to carry goods or push trolleys.

Staff turnover, if high, or temporary labour can introduce additional complexities. A budget must therefore be introduced in such circumstances for the issue of new credentials and system programming.

Software

The PC is an essential component in the operation of any business and the interface between the employee and the rest of the organization. Operators increasingly use the PC and software to control elements used in the access control and security industries and it is wrong to ignore the advantages and advances made with access control and security management software.

Commands are accessed through a range of iconized menus for point and click operation and administration will give access to the overall system control and programming to give a multitude of functions with comprehensive on-screen information. Real time monitoring with view of movements and alarms on building plans and maps can be achieved to interface with most reader types. The use of a PC-based system or the future introduction of one should be catered for at the design stage. See also Sections 4.3 and 6.4.

Integration and upgrading

The design should enable the access control system to be interfaced with other systems related to security or management. Equally it is important to establish if any future plans for expansion exist. This is part of the subject covered by software and recognizes that the system controller and network of cabling have a performance envelope and maximum capacity. In some instances integration can initially be through general inputs/outputs but more sophisticated signalling may be a future need.

The location of all of the readers should always be included on the design information chart together with the number of units employed at the controlled point. If roll call, anti-passback or time registration is to be included two readers are needed. The exact position can also be specified. The reader should be adjacent to the opening side of the door or barrier and it is to be appreciated that the majority of users prefer to address the reader with their right hand. The standard height from floor level is 110 cm but this can be altered if the reader is to be used by occupants in cars or goods vehicles or if the majority of entrants are in wheelchairs.

Controllers can be logged on the design information chart together with the position of any remote power supplies. The location of controllers is important together with the positioning of power supplies because there may be multiple units sited in areas that are not normally accessible.

5.3 Sensing devices

Although we have previously identified simple switches to monitor doors or barriers being closed or left open there is also a range of additional sensors that can be used to feed information regarding the condition of strategic points of the access control system. These detect a condition and report any changes back to the point of management of the system. We class these as inputs as the report is made into the system management hardware. These sensors or detectors, although not generally visible, can be an essential part of any system. Indeed these are often sensors that are also used in other security sectors.

In practice any sensor can be integrated with an access control technique even if a relay must be used as an interface. The use of sensors can include applications such as the monitoring of an area to ensure that there is no obstruction such as a vehicle before a barrier can be lowered or a gate closed to its required position. The type of sensor to be specified must of necessity be of the correct security level yet it may be active or passive.

Active sensors are so called because they introduce energy into a designated area. They can be integral units having a transmitter and receiver within a common housing. When the receiver notices a change in condition of the transmitted energy such as a beam being broken or a field of energy becoming distorted it will give an input to the management of the system.

Passive sensors are passive in the sense that they only measure a level of energy and do not transmit it. The most widely known is the passive infrared (PIR) which monitors infrared levels within its field of view and then activates if the change is greater than a prescribed threshold across the sensitive zones that it views.

Sensors may need to be energized by the regulated voltage of the system power supply or they can be connected to their own unique power supply which may have a standby battery support.

We can overview the most used sensor types and the technology that they utilize. In the main they have voltage-free contacts.

Audio

The receiver responds to an audible sound and can be set for sensitivity. The sensor can also amplify noise that can then be heard at a monitoring point. The audio detector may be sensitive to sound at a given frequency or be for use over a range of frequencies. They are not recommended in areas subject to high levels of background noise.

Capacitance

The capacitance sensor produces an electrical field between two charged antennas to form a dielectric space that is distorted if disturbed by motion of a body. The electrical resonance in the detector senses a change in current as an effect of the change in inductance or capacitance by a target entering the capacitance field and the system processor analyses the magnitude of change to assess the probability of cause. These sensors are for use indoors as they do not tolerate atmospheric changes. Often found in cable form applied on fence top posts, they carry a low voltage and use the surrounding air to store the charge.

Infrared

Passive

The passive device as used extensively in the intruder industry receives far infrared rays that it then focuses onto its pyroelectric sensor elements. These infrared rays are then absorbed and transferred into heat. As the amount of IR energy being received changes, the elements themselves alter in temperature and create an electrical signal. It is this heating and cooling of the pyroelectric element that is analysed by the detector. The risks of false activations can be reduced by using dual opposed or quad element PIRs or by combining the technology in a common housing with a further technique such as microwave. The latter detectors are called dual technology.

External PIR detectors use a similar technique but come complete with additional protection against the weather. The passive infrared has excellent detection capabilities of sensing human targets with a range generally in the order of 12 m in a volumetric pattern but with a range up to 25 m in alley pattern form.

Active

Active infrared sensors are essentially beam interruption detectors and can cope more readily with extremes of climate and temperature. They comprise of two main components, namely an infrared (IR) transmitter and receiver. When an object interrupts the signal between the transmitter and receiver the alarm output is energized.

These devices operate at wavelengths in the region of 900 nm at a carrier frequency of 500 Hz. They use an IR LED within a protective enclosure, which is filtered to operate at the correct frequency and pulsed rapidly to give a concentrated beam of IR that does not generate heat. The receiver is contained within a single chip using a photoelectric cell to transduce the energy to hold the alarm circuit in a quiescent state. The source is modulated to prevent the receiver being affected by another source. To stop the receiver responding to an incorrect IR signal, synchronization techniques are used to ensure that the transmitter can only operate with its correct receiver. The system is set so that the beam is monitored using time-based multiplexing.

Photobeam alignment is achieved by alignment status communication, the beam alignment level being visually displayed using LED indicators on both transmitter and receiver. The alignment status at the receiver is optically transferred to the transmitter with the lens adjusted for alignment using setting screws for vertical and horizontal levels. The operating range of active IR beams is generally in the order of 200 m, but in stable environments it can extend to 600 m.

Automatic gain control (AGC) adjusts the trigger level in response to environmental changes to cater for conditions such as fog, with heaters used in cold environments to clear the beam windows. The modern active IR sensor is extremely stable in operation as it uses multiple beam paths that must be broken simultaneously so debris or birds passing through the beams do not create a problem. The only conditions are that the active IR detector is not to be sited in an area subject to strong sunlight or from headlight glare along the transmitter/receiver line.

Light sensors/photoelectric

These are not unlike the active IR sensor as they are also classed as a beam interruption device. They use a transmitter and receiver, although both parts can at times be found in a common housing with a mirror or the object being used to reflect the tightly focused beam of light back to the receiver for analysis. Laser beam sensors have great beam strength and focusing capacities and are becoming more popular, although are more expensive than their standard counterparts. There are three designations of light sensor variants:

- Diffuse reflective. The transmitter and receiver are housed in the same body and sense an object by detecting reflection from it, and since all colours will reflect some light it can be used to detect almost all types of object. Although transparent objects can also be detected this version should not be used if the background is more reflective than the object. The range of this version is somewhat limited and is in the order of 0.2 m.
- Retroreflective. The transmitter and receiver are housed in the same body and the light beam is reflected off a prismatic reflector mirror. These devices are at an advantage in that they only need the installation of one component, but they are unsuitable for detecting objects that reflect such as polished metal surfaces. Their range is in the order of 13 m.

- Throughbeam. These types have a range from 2 m up to 50 m and comprise a separate transmitter and receiver. They use a photoelectric amplifier and come in a range of mounting configurations. The throughbeam is a true beam interruption sensor and being compact in size with good ingress protection it is used extensively in barrier systems.

Microwave

The microwave device used in external applications for beam interruption duty radiates its beams of microwave energy from a transmitter to a receiver that is in direct line of sight. The receiver compares the energy level with that transmitted and if a significant difference in waveform, amplitude or magnitude is found within a given time scale it goes into alarm. This is in contrast to the doppler effect or shift in frequency used by microwave detectors in clearly defined internal application areas.

The properties of microwave detectors are governed by the type of transmitter they feature. In general they have a greater depth of coverage than the IR or photoelectric sensor. A problem with the microwave detector is that their field can drift and they can also be affected by fog, rain and frost.

Switches (microswitch)

These sensors are an electromechanical device in that when a mechanical force is applied to its actuator its contacts will be moved. They are available in a variety of forms and may have plungers or levers. The microswitch also has the capacity to handle high electrical loads direct and the risk of false operation is almost non-existent if it is installed correctly as they have clearly defined operating characteristics. The microswitch will be found in roles where repeat accuracy and small movements are required.

The limit switch is a variant of the microswitch but is essentially larger and in a robust housing. It tends to be used in severe conditions, for example where it may be subject to the weather or when monitoring barrier positions as opposed to doors.

Switches (reed)

Magnetically operated the reed switch comprises an internal mechanism of two slender metal reeds sealed in a glass tube and positioned so that their ends overlap slightly but do not touch. Actuation is by placing a magnet in close proximity that causes the reeds to make physical and electrical contact by mutual attraction. The reed switch operates without the monitored parts having physical contact but has only a low electrical switching capacity so can only be used in signal low power circuits.

Switches (mercury tilt contacts)

This switch comprises an encapsulation that contains a pair of contacts bridged by a ball of mercury that is free to roll around the interior. The encapsulation is sealed and filled with an arc suppressing inert gas. The on/off state is dependent on the attitude of the switch relevant to gravity. It is used as a tip-over device.

We can conclude that it is possible to interface any recognized switch or sensor with an access control system to identify the condition of a component within the system. There are many detection devices and sensors used in the intruder sector that are inherently reliable, conform to BS 4737 and are intended for internal applications. Many of these use dual technology so that different recognition techniques are combined within a common housing and must respond simultaneously to generate a signal so that the probability of unwanted activations is low.

There are other external sensors in use such as fluid pressure, electromagnetic cable, fibre optic cable, geophones and piezoelectric detectors that can also signal the presence of a person or vehicle, but these would not tend to be used in an automated or high traffic application. In practice these are used as perimeter detection devices to check for unauthorized entry being attempted at a boundary or to start up CCTV equipment.

The capacity of the sensors to switch electrical circuits will vary and those that do not have volt-free contacts will come complete with their own control equipment.

5.4 Remote signalling

This can be performed to a central station or any other collection point anywhere throughout the developed world using the appropriate telephone network. As with any wiring system there will always be a number of regulations governing the installation of telephone wiring to prevent any damage to existing equipment or any injury to personnel using it. In practice each subscriber's line consists of a single pair of wires that connect the subscriber's premises to the local exchange. This may be called a direct line or DEL (often the American version is used where the exchange is called a central office (CO) and the line is referred to as a CO). The line carries the voice frequency communication in both directions and may be in the form of speech, modem or facsimile (fax) tones or other tones. This line also carries the signalling information with some signalling in DC and some in AC. When the call to the exchange is initiated the subscriber indicates to the exchange the number of the called party in one of two ways:

- By means of loop disconnect (LD) or pulse dialling effectively turning the current on and off. This is a development of the first form of automatic dialling and operates by temporarily disconnecting the local loop, once for the number one, twice for the number two, through to ten for the number zero. The timing of the pulses for the pulse width and time between successive pulses and the time between successive digits is controlled in order for dialling to be successful.
- By means of tone dialling or dual tone multifrequency (DTMF), which is the sending of musical tones for the number dialled without interrupting the connection. Tone dialling can only be accepted by more modern exchanges. Each digit is represented by two simultaneous tones as generated within the telephone and selected at the specific button on the telephone keypad. At the exchange the tones are decoded and the call route controlled. It is a faster and more reliable method of dialling than loop disconnect.

In practice the speed of dialling is extremely reliable as the decoding and switching are all performed electronically and the tones can also be used for signalling purposes after the connection has been made. The network operators have also introduced a further means of accessing the public switched telephone network (PSTN) by the use of digital techniques in that the customer can make use of the wider bandwidth and higher data transfer rates of main trunk transmission networks. This is namely the integrated services digital network (ISDN). The wide bandwidth allows the use of high speed data transmission plus high quality speech and video conferencing. The ISDN is divided into data channels of two different types and these are separate from the signalling channels. The data stream is therefore interrupted by control signals that enable the highest possible rate of data flow.

A consideration when using remote signalling is how secure the transmission must be as there are effectively four common lines all with different levels of security:

- The direct line (individual). This is a direct line not linked to an exchange and is used with a communication device that can constantly send a coded signal that monitors for line failure. It has high installation and maintenance costs and is only used in high security applications.

- The shared direct line. Similar to the individual direct line but also carries signals from other premises in the same general area. It is for mid-security risks, is more cost competitive and employs multiplex techniques carrying a number of signals simultaneously with the signals multiplexed onto the line.
- The indirect non-dedicated exchange line. This is a common method of connecting dialling equipment being one direction only but through an exchange. It must be ex-directory and outgoing only. It is not monitored and is used for low security risks and is best when underground or not accessible.
- The dedicated exchange line. Similar to the indirect non-dedicated exchange line but dedicated to the access signal only and not used for any other outgoing calls. It is not cost effective as the use of the line is restricted but gives reasonable security to the exchange.

The security level required must therefore be balanced against the cost budget for the signal transmission and the line available or the conditions attached to the installation of a new line.

Paknet

The radio data network may also be an option using Paknet. This technique is effected by high speed digital radio transmission through a network of national radio base stations and can be adapted to fit any system or control equipment. Using duplicated network structures the base stations have at least two separate paths from each location to the central station. The installation requires the fitting of a special interface device and a small antenna to send and receive signal information over the radio data network. On activation the interface instructs the integral radio pad to transmit a short message to the nearest Paknet base station through to the central station or final monitoring point.

The equipment is easily installed to interface with the access control system in the same fashion as the more traditional remote signalling equipment types of the speech dialler or digital communicator that use the hard wired telephone lines.

Speech dialler

This is a modern, inexpensive option of relaying messages to a number of personally programmed telephone numbers when the access control equipment generates a trigger. This can be the application or removal of a voltage or volt-free contacts changing over. The speech dialler uses a voice recording stored in a battery backed RAM and may be mains wired with battery support or be driven by the access control equipment's regulated supply. These may be actively promoted but it is important to recognize that they are of low security.

Digital communicator

These are prominent for mid-security risk signalling. They can be add-on or plug-in modules and consist of an NVM holding the telephone numbers to be called together with the client's identification number. They feature selected channels on the PCB to form a code on the central station receiver to identify the channel on the communicator that is generating the signal. This is sent as a series of pulses formed into a binary code and into a tone code by the modem on the communicator before transmission occurs. On receipt at the central station the tones are reconstituted into an alphanumeric printout or VDU format.

The digital communicator varies enormously in capabilities and can come in the form of a digimodem to enable upload and download to be achieved remotely from a PC. The essential part of the installation lies in the correct programming of the communication device and its functions and channel start/inputs so that the interrogation at the receiving centre is not corrupted as the range of functions and parameters varies considerably.

The connection of the line is common:

5	A	WHITE/blue	BT line
2	B	BLUE/white	BT line
3	C	ORANGE/white	bell suppresion
	E	electrical earth	

In the event that high security is wanted RedCARE can be invoked.

Communicating Alarm Response Equipment (RedCARE)

This is a link between the access control system and the telephone exchange that is monitored at all times so if a signal is sent or the line is cut or lost it is recognized. It also works if the line is engaged so there is no need for a new line unless it is also used for data transmission by fax or modem, for example.

The connection at the access system is by a subscriber's terminal unit (STU) to a premises alarm communication equipment (PACE) transmitter which sends its signals to a scanner at the telephone exchange and then onto a host at the CARE network control point.

Emergency and fault signals are sent through two routes so even in the event of a network fault there is nothing to prevent the signal from being received. Signals are encoded in a unique way and are only recognized by the RedCARE equipment with the system offering up to eight channels for the different functions and parameters.

RedCARE is seen in the industry as a very secure transmission medium because of its monitoring technique. Dualcom, however, can be considered as an alternative method by using a combination of Paknet and a digital communicator so that Paknet monitors the standard telephone line for signal failure or line cuts while the digital communicator monitors the Paknet link. Each service essentially monitors the other.

Although these signalling methods may be unnecessary for many access control systems they can be found in some integrated systems of which access control plays a major role.

Point ID reporting is also used in integrated systems and intruder detection as second zone reporting to verify an activity as having taken place.

Alongside the need to realize the role of these innovative communication techniques is the need to understand how uploading and downloading can be implemented. Systems that send a signal to private central stations are well known within the industry and are used for many applications. The growth of such facilities is a testimony to the advantages of remote signalling. This transmission optimizes resources and enables the operators at the monitoring station to interrogate systems and make valued judgements.

Up- and down-loading of an access control system is the process of allowing it to be remotely programmed or read via a standard telephone line and a PC. The ability to program the access control system from a remote location rather than at the installation itself offers tremendous advantages to system users and to the installer. Even though the central station will have the necessary programming capacity to perform this function even the mid capacity PC can carry out the task by connection to a modem.

For the end user the downloading protocol offers, above all, peace of mind. If system requirements change, the parameters can be easily reprogrammed at any stage to allow for additional users, change of access levels and time schedules plus voiding of tokens and credentials. It is therefore possible to define uploading and downloading as enhancing communications. It must be understood that a digital communicator will only allow basic alarm status with activations to be transferred to a dedicated central station whereas the digimodem permits the full status to be transferred plus the programming of all of the protocols together with upload and download.

As an alternative, software can be installed on a portable PC with a PC interface (PCI) allowing the data to be uploaded or downloaded from the control panel on site.

Independent software modules can be adopted for various security management applications for sites with multiple buildings. In this case the operation is controlled by the use of detailed graphical representations with map and site references. Monitoring software further enables the interrogation of various sites to be carried out by a head office so that all its local sites can be overviewed at any point in time.

The extent to which remote signalling is applied very much depends on needs and budgets but even by using basic signalling devices and standard lines access control systems can be customized.

5.5 Integrated systems

A fully integrated system can control and monitor all of the conditions of a number of security functions and keep the historical data in computer files. It can therefore manage complex security requirements and such systems often use components that are from different manufacturers but are brought together as a package.

With an integrated system the control is performed through software supervised by a single computer. For the system to be fully integrated all the systems must share a single controller and software with all the connected devices generating records that are held in clearly related files. In reality any security system can be integrated alongside the access control function.

We can say that integrated systems are brought about by the art of combining diverse elements into one whole or collective central control. Access control is often combined with CCTV but it is possible to go well beyond that level. The joining of different systems can allow the sharing of information so that integration can be seen as a system that covers all possible security as well as building control applications. It can be as simple as connecting an access control system to an intruder panel to stop personnel entering an area when that area is armed and no access is allowed.

Access control system ═══════ Intruder panel
↓
A pair of wires from the intruder alarm output to the door controller that secures access for that area

Alternatively it can be as complex as having a single point of administration for all of the systems in a building.

System X ——┬—— System Y

Software created so that the user interface reflects all systems that exist and how they have been installed

There are a number of reasons for integration:

- Tailored systems can be achieved.
- One network.
- There is a single administration point.

There are considerations regarding integration as existing systems may look similar with Microsoft Windows-based software so they have the language and understanding of the facilities. However, in practice they each only control a part of the building or security functions so they need to talk to each other to control the security as a whole. It is difficult to have existing systems talk to each other and this is always best designed in at the initial product specification point. Even with systems that have similar graphical user interfaces, extra intelligence must be added between the systems to translate the messages from one into actions in the other.

If a failure occurs it may be difficult with one integrated system to determine which system actually has the fault. If special software is linking two systems it is advisable that the systems can also be managed separately with individual maintenance. The features that can be accessed via the integration interface must be capable of management from outside of the core system, with some features only accessible from the core system to maintain reliability.

It is appropriate to say that the options on integrating access control systems with other systems are:

- Proceed only with discrete systems using simple interconnections via standard electrical input/outputs.
- Use sophisticated systems with bespoke software that has been 'designed in' at the outset.
- Have simple interfaces, allowing installation by non-specialist companies.
- Seek products specifically intended for integration.
- Use intelligent gateways from product to product.
- Ensure that the products can be commissioned by the access control installer and not by specialists.

Access control and CCTV remains a popular integrated technique and simple integration can be achieved by NO or NC switching, allowing one system to control the other by a pair of wires. The access control system can alarm when user specified conditions occur and the CCTV system can be programmed to invoke a particular camera and display the video as the access system alarms.

For facilities needing greater control of the CCTV system it is normal to use a CCTV controller with access control integration. This enables high levels of multiplexer control and switching capabilities allied to tracking by cameras. This is achieved by the CCTV host port opening communication between an access control smart panel and the CCTV monitoring equipment.

The market for the integration of systems is concerned with several issues such as technology, installation, operation and maintenance. This has all been brought about because a structured planning and management approach to security aspects has led to integrated systems and therefore intelligent systems.

We have therefore become aware that any one form of equipment does not need to be standalone and can be used more effectively if operated in conjunction with other components. However, it is the engineer's responsibility to consider each site as a separate entity and not to follow trends from other premises. It is only then that a truly integrated system can be an effective approach to a complex security requirement. The risks must be correctly assessed and satisfied if the integrated solution for a particular site is to be truly effective. The essential steps in the survey are therefore assessment of the risk, correct positioning of components and the installation. It also becomes apparent that the difference between installing an access control system and a system integrated with others is enormous if the hardware is to be linked into a bespoke system. Nevertheless it will become more common to use highly intelligent software in integrated systems to involve a networked system that controls the hardware either locally or remotely.

We can conclude that system integration can be easier if we adopt software as a common link to hardware but this may be difficult if the industries have more than one protocol for communications between equipment. Unfortunately we have industry standard protocols for communication between components of electronic equipment and others for procedural software. There are also other protocols for individual product lines. It follows that simple equipment with an ability to switch relays can be used for many applications. It must also be said that for a fully unified system with all of the individual systems controlled from a central computer to be highly effective, there must be total harmony of hardware and software using products designed for integration and with bespoke software 'designed in' for that exact system.

To this end we are now seeing integrated security control and management systems being introduced that can be used for a small private house but can be expanded to cover large industrial and commercial sites. These are purchased as a compact controller that uses a bus architecture for detectors and control devices so that 'tapping' into the bus provides for easy and less disruptive expansion. The zones cover fire, personal alarm and intruder with an ability to read a number of tags for access control duty. Additional applications are call with mimic display, CCTV switching, lighting control, hold-up alarm and deterrent warning. These controllers have built-in diagnostic facilities with volt/ current/resistance meters for electrical service. The essential access control options of adjustable timers, request to exit, heavy duty relays and full reporting are catered for in these systems to include multiple zone groups controlled by readers.

Progression in security will outdate standalone systems and advances will be made for integration to be effected from the small residential property up to the large site complex and equipment will be available for all of the duties at the different levels. This will be equipment designed for the integration of systems so it will be simple to install, use, expand and service. Allied to this will be a capacity for communications and remote signalling with service history and reporting. It should be recalled that the survey must be based on this concept.

The future holds great potential for the previously mentioned integrated security control and management systems that are effectively purchased as a kit. As an example of how they are applied we can overview their basic cabling. The actual wiring is relatively straightforward and at the access door station printed circuit board there are typically four connection blocks each holding several terminals identified by numbers.

A connection block. Wired to the system RS.485 bus.

1. OV   2. +12v   3. A   4. B

B connection block. Outputs etc.

1. Lock OV   2. Lock +12V   3. Lock NC   4. Lock C   5. Lock NO
6. Supply OV   7. Supply +12V   8. Alarm NC   9. Alarm C   10. Alarm NO

C connection block. Wiring to RTE and monitor.

11/12. Exit button   13/14. Door monitor

D connection block. Wiring to reader head.

1. Black (OV)   2. Red (+12V)   3. Brown   4. Blue   5. Green   6. White

These connections and terminal numbers are similarly identified at the termination points.

On the printed circuit board there is a connector to plug onto the intelligent power supply, tamper protection switching and appropriate fuses, but supply to the lock mechanism is to be derived from the local power supply. Certain traditional notes are made:

Keep the access reader cable separate from the lock supply connections and use twisted pair to the reader head with termination of the screen to the main earth at the door station only.

For DC lock mechanisms incorporate back emf protection diodes as specified by the manufacturer.

Lock supply connections and alarm output wiring through the volt-free relay contacts follow standard procedures and the addressing is via coded switches mounted on the same door station printed circuit board.

This door addressing selects:

Lock open time: Length of time lock release will be activated

Door open time: Length of time door may be held open before a local alarm is generated

Door contact number: A zone number valid for the integrated system

The access system as described uses proximity tags that are themselves programmed using an LCD keypad through the Manager menu at the main system end station printed circuit board.

5.6 Site detection and monitoring

To conclude this chapter we investigate the techniques that exist in order to detect attempts by unauthorized persons to circumnavigate the external fencing or barriers in an effort to avoid the locking devices and barriers. There are technologies available to us to enable security personnel to observe and monitor the progress of these people as they move around the site.

There is available a good selection of perimeter fencing that can be adapted to deter and physically restrict unauthorized access to an area but this can always be supplemented with electronic detection to integrate with the access control system. In the main this integration invokes CCTV monitoring so that any detection event is verified. We have previously looked at sensing devices that are used to monitor for doors or barriers being closed or left open and to feed information back to the system with regard to obstructions being in the movement

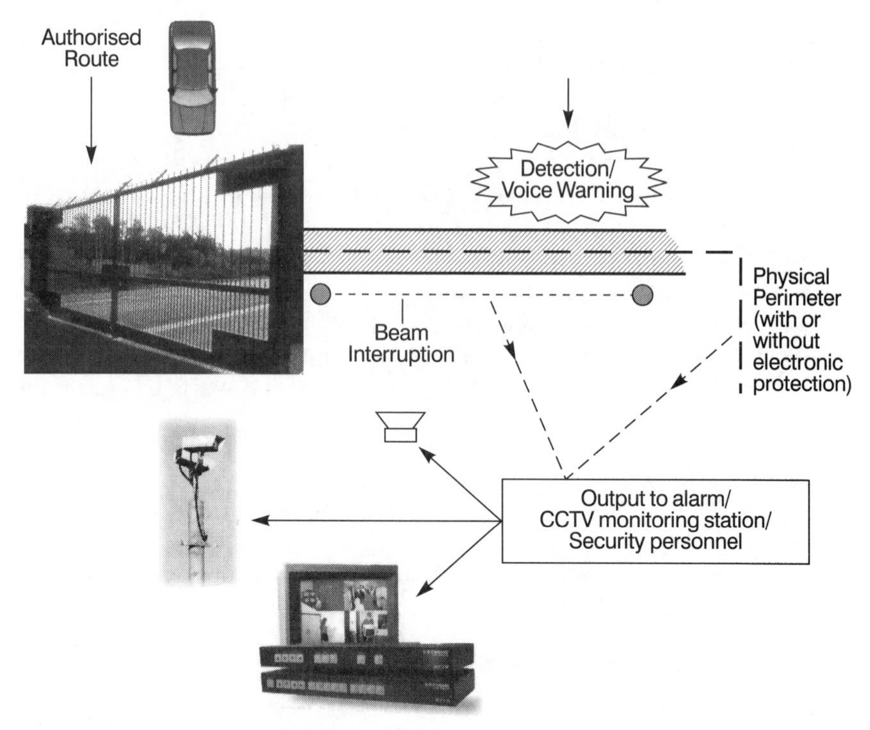

Figure 5.4 *Site detection and monitoring*

path of barriers and doors. There will always be a certain overlap in the techniques these sensors apply and those that can be used to monitor fences for attempts to bridge or cut through them. There will also be certain beam detection units that can be employed for simple monitoring of gates and for long distance detection of an intruder having cheated a barrier and gaining entry to a restricted area.

In the first instance we shall overview the protection techniques used with fences, walls and trenches. In all cases these detection types do have considerations with respect to unwanted activations due to the very nature of their use for perimeter protection. They may also be activated quite innocently by passers-by. For this reason the activation of the sensor should be verified either by a dual technique or visual practice even though all the sensor types do have a method of adjusting levels of sensitivity.

The architecture to be considered in this section is shown in Figure 5.4.

Fence, wall and trench protection

Fluid pressure

This sensor type comprises a small diameter tube sealed at one end and with a flexible wall that if compressed causes a pressure to be applied to an enclosed fluid. The change is analysed for magnitude and the energy is converted to an electrical pulse. The operation of the sensor is caused by a person or object applying pressure upon the ground in which the tube is sited.

In practice the tube may consist of a number of loops and multiple loops can be analysed for pressure changes over a large expanse close to a fence or wall. The tubes are located in trenches in accordance with the manufacturer's data but are not to be used under surfaces that can have an elastic effect such as gravel, sand, asphalt or grass. Solid surfaces such as concrete can cause complications and are to be avoided.

False alarm sources can be as a result of ground movement or from the foundations of objects flexing in the vicinity of the sensor during high winds.

Electromagnetic cable

These detectors consist of a cable with the inner and outer conductors separated by a dielectric filler. It operates in such a way as that when a current is applied to the separated conductors an electromagnetic capacitive field is created but this is interrupted by the dielectric layer. The change in the outer conductor is caused by external pressure being applied and this mechanical motion is transformed to an electrical signal. The amplitude of the electrical signal pulse is in proportion to the external applied force.

These sensors can respond to vibration, cutting or indeed most forms of interference.

In general electromagnetic cables are fixed to the inner surface of a fence with the fixings related to the particular cable while greater level of efficiency is afforded by a greater number of fixings. Often these devices are classed alongside microphones and can be found buried in gravel.

False alarm sources in these cases are governed by the sensitivity setting of the processor and the environmental disturbances that can be found.

Capacitive sensors

These use electrical resonance to sense a change in inductance or capacitance by a person entering the detector's capacitance field or they will respond to touching of the resonant tuned circuit.

They can be found in a number of different guises. In cable form they can be applied to the top of fences or walls and by carrying a low voltage they provide a capacitive effect between a grounded post and the wire. They may also be found to use the surrounding air to store a charge.

The system processor analyses the magnitude of change to assess the probability of cause but false activations can be attributed to driving rain and snow.

Fibre optic cable

These are efficient as sensor cables but must be cut or badly damaged to create a signal. They are intrinsically safe so can be used in hazardous areas as they are incapable of causing dangerous levels of power that can ignite flammable gases. The receiver of the transmitted signal responds to a change in magnitude of the received signal once this is removed or distorted by cutting or damaging the cable.

The cable is installed in a hollow core through which access may attempt to be made. The cables are not surface mounted as they are easily damaged.

The incidence of false activations is low and these sensors can be used over long distance and they do not become affected by interference or spurious electrical signals.

Geophones

These devices are made up from coils with a magnet held in such an orientation by a spring tension to create an electromagnetic force. If the magnet becomes displaced by a positive movement of the sensor, or by the influence of pressure, a current becomes generated in the coil and this is analysed for magnitude to determine if it is an alarm source.

Geophones can be installed on fences and walls or any other object that could be subjected to cutting or sawing and they may also be buried.

Their ability not to false alarm is governed by the surface to which they are mounted not to vibrate. If they are to be buried this should not be in an area surrounded by trees.

Piezoelectric

These respond to mechanical stress, strain or compression producing a proportional voltage to the applied mechanical medium. The piezoelectric coefficient is its defined ability to determine to what degree the mechanical stress has been exerted.

These detection types are installed as a cable with the mechanical strain recognized via an analyser that is set for sensitivity. These can be buried or surface mounted.

The device is set for sensitivity depending on the environment in which it is to be applied.

The fence, wall and trench detection devices that have been covered can all be integrated with any access control system to monitor access through an area other than the barrier or gate and they will be found to differ from the competing manufacturers. The systems can appear under different trade names and be promoted as a means of detecting and deterring intrusion to provide enhanced automatic environmental control with a discrimination to eliminate false alarms.

As a stand-alone technology they can be retrofitted to most existing structures and applications and by means of a zoning principle the location of an attempted entry can be easily determined.

The installer may also wish to consider the use of electro-fences which are wires placed at the top of a wall of fence on sprung steel mountings that provide a deterrent to the placing of a ladder against the wires. The device carries a safe but powerful electric current and any attempt to cut through the fence or reduce the voltage immediately generates an alarm while a pre-defined timespan or a short circuiting of any of the live wires causes the same effect. Sprung steel supports increasingly flex as the intruder puts pressure on the wires hence causing the required activation.

All fence, wall and trench detection techniques are excellent at providing perimeter protection since they signal an intrusion at an early stage. Beam interruption detection devices also have a role to play and certain of these can actually be found on the top of walls and fences to stop any scaling of the perimeter. Alternatively they can be used inside of the protected area close to the fence or wall.

Beam interruption detectors

Active infra red

The most used beam interruption detector is the active infrared. This sensor did gain a mention in Section 5.5 as it may be used to monitor for doors or barriers being closed or left open. It may also be used in order to feed information of the status of an area back to the management point such as that area through which a barrier moves through being left clear of obstructions. However, they are also widely adopted to monitor for intrusion in a protected area to ensure that a barrier has not been cheated or to stop a wall or fence being scaled. The latter is achieved by directing the beams along the top of the wall or fence. There are a number of additional factors that are to be addressed when the active infra red is used as a detection device as opposed to monitoring a smaller closely defined area and it is these we now consider.

The active infrared is ideal as a perimeter sensor in detecting intrusion at an early stage because it can not easily be masked and since it monitors the entire length of its coverage area it can produce an output even if one of its multiple beams is blocked. This actually prevents the system being set until the beam is cleared and we may say that false alarms are avoided because of the use of these multiple beams and the fact that once set they must both be broken simultaneously for the alarm relay to operate.

The following factors should be considered:

Bad visibility IR at 900 nm operates quite well through fog, but in areas where this occurs on a regular basis it is necessary to use a sensor with automatic gain control. This continually monitors for gradual changes in the signals caused by variations in the weather conditions. It then adjusts the trigger level accordingly to maintain the proper sensitivity level for the existing environmental conditions. In fine weather the automatic gain control keeps a high trigger level to prevent interference from external light. In bad weather conditions this circuit automatically lowers the trigger level as the IR energy becomes blocked by rain, snow or fog. In the event that the signal is lost an output is generated as 'trouble' to advise of detector closedown.

Air turbulence This is a greater problem in hot climates and long range applications, and it is necessary to reduce the operational range to compensate.

Low temperatures Anti-frost designed slots on the cover can allow the beam energy to pass even when the cover is completely frosted over. If the device has a high tolerance to signal loss, then even a small amount of IR energy will ensure stable operation. Beam systems can also feature heaters and thermostats to clear the beam windows from frost and ice. Heaters may be included in the optical head of the beam sensor itself. They can be placed within a tube to make use of the vortex effect. This design accelerates the flow of air over the heater and projects warm air in the position where the beam passes through the outer cover. When used with towers, one thermostat is needed for each tower and should be fitted at the bottom.

Power supplies These may need upgrading when heaters are used, and can be placed at a number of points in the larger systems and close to the sensors.

Anti-tamper The system must be fully tamper protected, and when towers are used a top tamper switch should be incorporated to prevent the tower being used as a climbing aid so that downward pressure generates an alarm signal.

Strong light The strong light from the sun and headlights that can shine directly onto the transmitter and receiver are to be avoided. If strong light stays in the optical axis for a long period of time it will affect the life expectancy of the sensor.

The other considerations are that the units should not be installed in an area where splashing dirty water can be encountered. Maintenance should include checks for no obstructions being present between the transmitter and receiver and that the ground on which the devices are mounted has not become unsteady.

IR beams are governed by BS 4737: Part 3: Section 3.12: 1978, 'Specification for components. Beam interruption detectors'

Microwave

This is the only other beam interruption detector that is used in volume to compete with the active IR but is not as popular although it is less affected by changes in the weather.

The microwave detector is, however, a highly sensitive device affected by moving grass, bushes or trees and is difficult to commission as it is not easy to determine the actual range being covered. The properties of the microwave beams are governed by the type of transmitter specified, although in general they do have a far greater depth of coverage than the IR beam but the field can drift and they are best used close to a boundary wall or fence only.

The microwave sensor can provide a secure area up to 200 m although it is best on long straight boundaries. In cases where there are ground undulations and there is a concern that an intruder can crawl under the beams, beam shaping or phase sensing is applied. This process uses vertical or horizontal aerials to give increased ground cover or extended ground and high level cover. The method of installation of these aerials varies with the manufacturer of the equipment.

The type of transmitter used will depend on the area that is to be protected. Vertical patterns have an energy form based on a parabolic trough that is very broad and gives protection against crawling. Although pole mounted, the sensors are only sited some one metre or so above ground level. Short-range detectors are used to cover small areas volumetrically so their beam pattern is essentially cylindrical. Long range beams can actually claim some 300 m distance but this is at the expense of beam width and it is more common to install several mid-range sensors.

These devices are governed by BS 4737: Part 3: Section 3.4: 1978, 'Specification for components. Radiowave doppler detectors' and reference should also be made to the previously noted BS dealing with IR beam sensors which holds additional information.

If we continue on the theme of looking at outdoor protection as a means of ensuring that an access control system barrier cannot be circumnavigated the engineer can consider innovative ideas such as voice warning functions. These are multistabilized outdoor detectors with multiple detection patterns and sequentially confirmed alarms using passive infrared as opposed to active beams. These sensors provide reliable performance and can discriminate between large and small objects so that activation by small animals is negated.

These sensors can be used in conjunction with other sensors to create even greater stability. They have a voice-warning feature that is essentially a weatherproof speaker inside of the device. This is designed to deter a would-be intruder from continuing towards the protected area instead of following the correct procedure of attending an access control barrier/point that is monitored by CCTV or security personnel. The audible message warns to the effect that the area is protected and access can only be granted at a different official entry/exit location subject to credentials being presented. Second sensors of a different technique can then activate an output if they are triggered because the voice warning has been ignored and access is attempted at a restricted point.

At this stage the access control installation engineer can decide if there is a need to supplement the system with CCTV monitoring or recording. This is readily overviewed.

CCTV

In practice the system will cater for observation or continuous monitoring.

Cameras

- If it is intended to identify and track a target the camera should be colour.
- For observing and if the lighting conditions are poor monochrome/black and white may suffice since colour cameras require more operational light than monochrome.

The security coverage level determines the camera lens and these can be summarized:

- Small area coverage: use a wide angle lens e.g. 3–4 mm with a standard resolution camera.
- Medium area coverage: use a medium angle lens eg. 6–8 mm with a high resolution camera to resolve small detail in the background.

- Large area coverage: use a zoom lens to zoom into detail or view the wider field with a high resolution camera.
- Specific detail: this is to identify faces and cars or such. Ideally two cameras should be used to resolve the detail at distance. It is best to use a telephoto lens 12 mm plus. Also a wide angle.
- Blind spots: these tend to be under the camera. The more narrow the lens angle the larger is the blind spot. In these situations use a wide-angle lens and site the camera as far as practical from the target and horizontal.

Although low voltage cameras are often used internally it is more normal to use mains voltage cameras for external duty. This depends on the length of cable runs and a d.c. voltage will always suffer voltage drop along the run. A 24 V a.c. system can be used if necessary with the power supply close to the control equipment for easy service and maintenance.

For the smaller application the installer can consider using combined cameras and detectors using a common supply.

Telemetry

If remote control of the cameras is needed telemetry should be specified. This offers a practical system in order to control the operation of equipment at a remote point. It allows the control of pan, tilt and zoom functions and to switch on lighting with movements to pre-set positions. The controller at the operating position is the transmitter with the receiver at the remote location. The signal can be sent along a twisted pair cable or the same co-axial cable that is used to carry the video signal.

New telemetry systems are touch screen computer based or have menu driven consoles offering real text descriptions and site plans. They modernize traditional joystick camera control as they are easier to use and quicker to respond. The architecture is as Figure 5.5.

Telemetry uses a man–machine interface (MMI) to drive the functions but a further technology to be understood is video motion detection (VMD) which is only intended to capture images when an unusual event occurs. It will then initiate an alarm response.

Video motion detection

In analogue form VMD responds to movement captured by a camera using a comparative analysis of pre-selected image areas between sequential frames which enables the detection of any change in illumination level greater than the system's pre-set sensitivity level. This

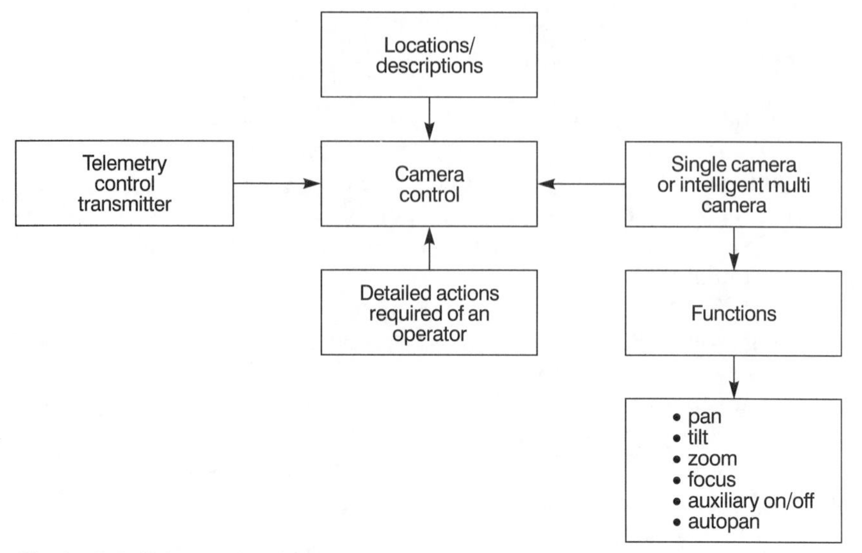

Figure 5.5 *Telemetry architecture*

may be a little troublesome in outdoor applications because of clouds, shadows, wind-blown shrubs and small animals.

False alarms can be negated in digital form as it divides an image into a number of detection parameters. Subsequent images are processed by the system and these are compared to pre-determined criteria. In view that the changes in the image as detected must conform with the criteria laid down by these settings the rate of false activations are extremely low.

Following a valid alarm condition, digital recording of all images in the area will commence. Each camera in a system can be programmed to record a variable number of pre- and post-event images to show video frames up to and including an alarm trigger. These can then make up a full video of the incident.

VMD is often used with physical barriers using 'double knock' as an alarm is triggered by first strike detection with the follow up VMD picking up the resultant intruder movement. Tracking VMD devices for use with pan, tilt and zoom (PTZ) cameras can track and isolate an unauthorized person's image. They may be an individual unit or be part of a combination system. The architecture is as Figure 5.6.

Multiplexer

The multiplexer is sited at the control point and it enables the surveillance system to capture multiple source images. Although switchers are still

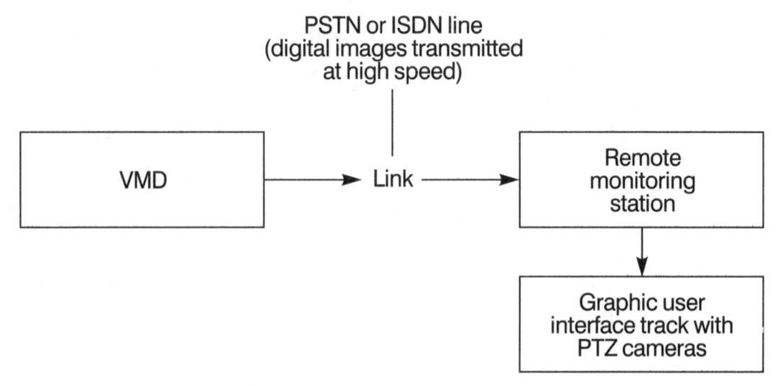

Figure 5.6 *Video motion detection*

used in some installations to switch between cameras for viewing purposes the multiplexer is now the preferred device.

These devices take the outputs from a number of cameras cutting them into time slices and then sequentially recording each segment. If the output from a certain camera becomes more significant such as if an alarm is generated on it the multiplexer will give that camera priority. This is achieved by recording just that camera or increasing its time allocation so that it is recorded more frequently than the others in the sequence.

The majority of multiplexers are controlled manually with the menus displayed on the monitor. The recorded images can be played in a variety of formats including one camera full screen, quad (with four camera images displayed) or picture-in-picture. A simplex system will perform just one of its key functions – record, playback or live multidisplay screen at any one time. A duplex system however can perform two functions, viewing live scenes while also multiplex recording.

Simplex cameras are the best choice when recording only is the priority. It can not display multiple images, i.e. quad or split screen while recording.

Duplex should be used if monitoring is being performed as it can provide screen splits and user selectable images without affecting what is recorded on the connected VCR.

We can say that multiplexers allow the recording of several cameras on one video tape. The multiplexer is intended for use with time lapse video recorders that can extend the recording time on a standard video tape for up to 960 hours.

VCR

These must also be addressed in order to produce recorded pictures of evidential quality. Although VHS can produce good images, S-VHS can increase resolution by 60 per cent minimum and are compatible with high resolution colour cameras.

The most popular time-lapse VCR is the 24 hour machine. For recordings of longer than 24 hours machines with time lapse capacities from 72 to 960 hours can be used. These all exploit the understanding that normal video recording produces 50 fields/s, most of which are identical. The time-lapse VCR extends the recording time on a standard video tape and as an example by taking one field from each camera it is possible to monitor 50 cameras in one second at normal speed (3 hour mode).

Allied to all of the foregoing equipment is a need to choose the best monitor for size and quality and to ensure that lighting levels for the cameras are adequate with particular attention given to colour cameras as these are intended to work across the full white light range.

The addition of CCTV to an access control system to give remote surveillance is a huge advantage in terms of security but must be balanced against cost budgets although in real terms there does exist a diverse range of equipment in various price parishes.

Live remote camera surveillance can be achieved digitally transmitted to a conventional TV, monitor or video recorder using a standard telephone line. These systems consist of two units. A transmitter is installed at the remote site where the requisite surveillance cameras and alarm detectors are positioned. The unit compresses video information into a format which can be passed down the conventional telephone line via an integrated, standard approved modem which is linked to the receiver at the host site. This receiver then enables the live video surveillance of the remote site to be viewed or recorded on a standard monitor. The images can be colour or monochrome and latest alarm situations can be stored and shown prior to real time images.

In cases where telephone based systems, either PSTN or ISDN do not offer an established communications infrastructure microwave links can be considered. These use microwave technology as the transmission path as opposed to hard wiring cables or a telephone line. They have a range in the order of 1 km.

For this installation type the cameras that are installed at the remote location are linked by modular plugs to a microwave receiver. This unit is interrogated using microwave technology by means of a transmitter located at the monitoring point. This checks for the video signal. The only consideration is that of aligning the path between transmitter and receiver. These microwave links are cost effective and easy to install and can be considered when cabling is not an option. They require little additional equipment other than appropriate power supplies.

5.7 Auxiliary components and duties

It is important that before the engineer moves to the installation and commissioning stage future duties of the access control system can be catered for if a need arises. However it is vital that the basic system does account in the first instance for the essential needs. On many occasions a system is installed at a base level but with the understanding that as budgets allow and as an area or site expands add on devices can be included under an expansion program. It remains to say that the surveyor must not overlook the essential components before the installation is put in hand as mistakes can be both difficult and expensive to correct at the later stages.

There are many auxiliary devices, both physical and electronic, that are easily installed within any system even at much later points in time but others involve certain considerations. The correct cabling at the first stage is important as this is difficult to rectify plus the controller and hardware because these very much govern the latitude of allowing any future expansion.

Throughout the course of this book we have broached many additional components and their role but it is correct to cover certain versions in their own right because they can also provide an idea of how systems can be upgraded with little effort. It is to be recognized also that many controllers and the main system hardware often have features that will not be used but the engineer has a responsibility to appreciate all of the options that are available and how these can be implemented as the need arises.

It will be appreciated that product lines will vary between manufacturers but we can record certain accepted components that have widespread and international use.

Mains filter

These can be used with any system and should be installed at the outset. They contain an effective RF filter to remove induced radio frequency interference and incorporate varistors to protect against high voltage transients that can be present on the mains supply causing microprocessors to crash.

This interference can come from a number of sources and cannot be identified by normal means but examples include lightning strikes, passing high powered radio transmitters, the switching of highly inductive loads and badly suppressed electric motors.

In operation if the transient spike is of a high enough voltage it causes breakdown of the varistors so the transient becomes safely earthed. The varistor once operated in this way must be replaced whilst the earth leakage current flowing can be protected by the RCD at its source so that the access control system becomes isolated.

These devices are not to be checked with high resistance testers such as Meggers as part of a system cabling test but must be disconnected.

They are installed as near as possible to the equipment being protected and can actually form part of the system spur. They must be well earthed and be correctly fused for the load.

In practice they become the interface between the system mains supply and the equipment supply connection point.

Relays

These are exceptionally versatile and available in many forms. They are easy to fit wherever voltage free switching is required or when it is wanted to switch higher than usual currents and voltages. Compact in size they can often be incorporated inside of the existing control equipment. Common applications include the switching of extra sounders and lights, the actuating of barriers or doors or the switching of surveillance equipment. There are a number of general observations that can be made with respect to their use in access control:

Polarized versions should be specified if the polarity of outputs is reversed in alarm such as when access systems are integrated with the switching of fire alarm sounders. These tend to have protection diodes fitted across the coil that of itself can be energized by a number of different voltages.

Double pole changeover relays can be used to control two circuits at the same point in time but with the switching of the circuits being isolated from each other.

Transistorized versions can be used to allow low current signalling outputs from the control equipment to actuate the relay without overloading the signalling circuits. Trigger voltages can be selected by a link and be negative or positive. They will have a general range from 5 volts to 15 volts with a current drain in the order of 4 mA.

Toggled relays give a latching on/off action that is activated by the trigger signal so that the first input turns it on and the second input turns it off. These are ideal for use with radio transmitters/receivers to operate doors, barriers and gates. Typical options are:

Positive trigger signal
Positive hold-off dropping to O volts (neg)
Negative 0V signal (resistor fitted across 12V and trigger)

Timers

Often used in conjunction with relays these devices are equally versatile with numerous roles in electronic access control.

Service timer

These are long term duration modules that operate a relay after a selected period in the order of 3, 6 or 12 months. They are intended to indicate to a subscriber that a preventative maintenance visit is required by signalling the control equipment to respond with a 'Call engineer' indication. Alternatively they can bring on a flashing LED or buzzer.

They tend to be time set by DIL switches with an onboard LED confirming that the timer is timing out. These modules need to be permanently energized and require battery support if the control equipment powering the timer can switch off the timer in the event of mains failure. The alternative is to use a unique power supply with its own rechargeable support battery.

Standard cut-off interface timer

This device removes supply voltages to component loads after prescribed time intervals. When an input voltage is applied the timer commences and then when it times out the output voltage is removed. It is automatically reset when the input is removed. They are effectively an interface between an input and an output voltage to ensure that the output must not stay active beyond a given time.

Modular timer

Many different modes of operation can be offered by this variant. The timer starts when inputs are applied as a pulse or a continuous signal. The ranges are usually selected by potentiometers and the modes by plug in links. Typical options are:

Mode A. The relay is on after the timed period.
Option i) Relay latched on.
Option ii) Relay activates momentarily. On for some 2–3 seconds. The relay will stay energized after time out if a trigger input remains present. The relay will de-energize and the timer reset when the input is removed.

Mode B. The relay is on for the timed period and then off. The relay will de-energize after time out even if a trigger input remains present. It will reset after removal of the input.

Digital timer. A further timer with generally two modes:

Trigger input – Timed period – Power on.
Trigger input – Power on – Timed period – Power off.

The modes are selected by links and the time periods by resistors. These devices switch an input to an output after a prescribed period of time but for heavier currents and voltages to be selected as an output they are to be used in conjunction with a relay.

Delayed release module

These are not unlike timers in working as an interface and providing a voltage free output. In electronic access control systems they are often used for door open warnings or to power door releases for an extended time. They are classed as delayed release because in addition to a latch facility they can be selected to give a hold on until the pre-set time delay has expired. The modes are typically:

Latch on after removal of the trigger input until reset.
Hold on until a pre-set time time delay has elapsed (Delayed release). This is adjustable.
Transistorized relay. This follows the trigger input.

The trigger inputs can generally be:
–ve start 0 volts applied.
+ve start +5 volts to +15 volts applied.

The latched mode can be set by shorting the reset pins or by a push button.

The timed mode will hold on the relay until the timer has timed out. This is adjustable and the relay will switch on and off as the trigger is applied.

These devices need a permanent supply so are powered from the control equipment with battery support or from a unique power supply with standby power.

All access control systems that come supplied as a product group or kit, complete with all of the components needed to make up a base system, will have a variety of available options. Standard examples are door panels or readers in different materials to withstand the various environmental conditions to which they may be subject and to include vandal resistance.

Spy shields can be installed for coded keypads so that the entry of PINs can be hidden and tone and LED indications are a requirement. Additional sounders can be fitted to give bleeps for partially sighted people and multiple door interlocks will allow groups of doors to be interlocked together. Printers can be linked to allow the print-outs of transactions and alarms. Lock outputs can signal further LEDs with buzzer outputs and additional LEDS can be used to indicate extra functions such as duress/panic alarm with perhaps also remote signalling. Timing options with system shutdowns can also be incorporated.

It remains to say that the components to customise any system to bring in extra activities can always be achieved by add-on components both electronic and physical.

It is the responsibility of the surveyor to recognize the extent to which these can be applied so that the system that is to be installed and commissioned is as far as practicable correct for its intended function. This must satisfy the needs of the client yet it must not be so difficult to operate that it becomes unduly complex.

5.8 Discussion points

The survey and design for the proposed electronic access control system must clearly cater for future trends, modifications, site changes and integration.

Discussions must also take into account the use of remote signalling options and additional sensors and detection equipment that could be used to start up equipment such as CCTV monitors or recording of images.

To this end it is vital that the engineer surveys for controllers and ancillary equipment that are capable of expansion within the reasonable expected limits of the future site and system development.

6 Installation and commissioning

In this chapter we address the procedures that are involved in the installation process of an electronic access control system. This involves the installation and interconnecting of the cabling and components plus the siting of the main equipment and is followed by the tests that are required at the commissioning stage.

An important part of the installation is the containment or protection of the cables plus the fixing to ensure their security and resistance to the elements, abuse and mechanical impact. There is also a need for an inspection of the equipment and associated parts plus the programming of the software. This programming may be performed at the controller or it may be necessary to carry out the procedure at a PC.

At the conclusion of the installation all the data can be logged at the commissioning stage in preparation for the handover to the client forming the concept of customer care.

A further part of this chapter is devoted to guiding the reader towards the Health and Safety measures that form a part of any installation.

6.1 Installation of system cabling

Regulations cover the segregation of cables of different categories and good practice is required to ensure the neatness of the finished system and protection of vulnerable wiring.

The routes that the access control system cables follow, the selection of these cables in accordance with the manufacturer's data and their method of fixing and containment form a vital part of the installation process.

The constructional features of the wiring conductors differ throughout the access control industry and there are also a number of termination and connection techniques in use. These terminations will have differing mechanical and electrical properties.

Cables will be found alongside flexible cords and some will have particular resistance to the environmental conditions in a particular area.

Reliability and ease of installation together with cost must be part of the terms applied against the installation of the system cabling. It is important to understand if there are any adverse environmental conditions at the site as any extreme conditions or potential causes of mechanical damage will reduce the level of reliability or security if not considered. Equally the weather and heat can induce problems of dampness and corrosion leading to deterioration of the cabling or faults within it.

Before we consider the requirements in detail we should recognize the basics needs:

- Where practicable, cables are to be installed within a controlled area.
- Where practicable, cables are to be concealed.
- If cables are exposed to possible mechanical damage or tampering or are in public areas, they should be protected by suitable conduit, trunking or armour.
- If an access point release signal passes outside of a controlled area, metal conduit or an equivalent containment should be used.
- All interconnecting wiring shall be supported and its installation is to conform to good working practice.
- Any cable joints shall be made in suitable junction boxes using either wrapped, soldered, crimped or screw terminals.
- Low voltage and signal cables are not to run in close proximity to mains or other transient carrying cables.
- Signal cables for the transmission of data or other low level signals shall be of a type and size compatible with the rate of data transfer and anticipated levels of electromagnetic interference.
- Cables shall be installed in accordance with the IEE Wiring Regulations for Electrical Installations.
- Low voltage cables from both mains and standby power supplies to remote equipment are to be of sufficient size to permit satisfactory operation of the equipment at the end of any proposed length of cable run.

Cable routes

The first stage is to become familiar with the site and to identify the routes used by existing cables and services. It may be possible to share these routes and yet still maintain segregation from the existing cables.

If this is not possible because the existing routes are no longer accessible, are sealed or have had all the useful space exhausted it becomes necessary to look for new runs and to establish the method of fixing the new cables. These cables may be fixed in position with supports or clips if installed in inaccessible areas or are not vulnerable to damage. However, if it is not possible to install these cables in protected positions it is necessary to consider containments to hold the wiring. These containments give a level of protection to the wiring and at the same time improve the appearance of the cabling installation.

In other applications the cables are simply fixed in position to support them and to stop sag. Running cables through voids is the preferred and most cost-effective method as it removes the need to install containments.

When running cables through voids it is important not to put strain on the cables and to allow for gradual bends otherwise the insulation can become damaged, which leads to a decrease in the insulation and performance characteristics.

It is normal practice to allow a cable bend to be of a radius in excess of four times the outside diameter of the cable.

Cable types

In all access control systems there is a need to run separate cables to the different components. The specific cable types to be installed are recommended by the equipment manufacturers and reference can be made to Section 4.5 for common versions.

In general cables to door sensors and locks will be unscreened but when used to connect readers and keypads they will be screened and of data transmission format.

Certain signals can be carried within different cores in the same cable without any interaction but data and control signals must be carried in separate cables and compartments. The cable types are therefore governed by a number of features including the compartment in which they are to be installed.

In the next section we discuss fixings and containments used to hold and protect wiring.

Other specific cable types are available that have integral protection and examples are steel wire-armoured cable and mineral-insulated copper-sheathed cable (MICS). The former has a braid of steel wire held under the outer jacket to withstand penetration and is often advocated for underground installations. The latter has solid copper conductors insulated with a highly compressed covering of magnesium oxide (MgO) powder and sheathed with a seamless malleable copper tube. MICS is extremely robust and does not need any further mechanical protection except in areas where damage is likely. It also has a higher current rating than other cables such as PVC or rubber insulated of the same conductor size because of the high thermal conductivity of the magnesium oxide insulation which allows the heat generated in the conductors to flow quickly from the cable.

Although wire-armoured cable and MICS are particularly durable PVC-sheathed cable of standard format will resist attack by most oils, solvents, acids and alkalis. In addition it is unaffected by the action of direct sunlight, is non-flammable and hence is suitable for a wide range of internal and external applications. PVC-sheathed cable can be run between floors and ceilings and dropped down through ceilings. Holes made for the passage of cables through ceilings can easily be filled with cement or another building material as a precaution against the spread of fire.

The IEE Wiring Regulations do not permit the running of extra low voltage cables with mains cables unless the insulation resistance of both is equal; however, it is good practice not to run access control signal cabling alongside any mains cables – neither should they be fed through the same holes in the building structure or passed through the same entry hole when cabling control equipment or a power supply.

The fixing and containment of the wiring forms a separate subject.

Fixings

Under this heading we should consider all methods of supports and containments. Reference should also be made to Section 4.6, which covers the ranges of cable protection.

The fixings we are interested in are:

- Steel conduit.
- Non-metallic conduit.
- Trunking and flange trays.
- Aluminium tubing.
- Capping/channelling.

Steel conduit

Although this has great strength the quality and nature of the assembly are vital if the installation is to have continuity and be aesthetically correct. Initially the installer must determine the number of cables that are to be drawn into the conduit as this determines the size of conduit to be used and its fittings. Equally conduit is classed as one compartment and for that reason cables of different voltages should not share the same run, nor should data and control signals. It is possible that the mains supply cable is carried in steel conduit but this should not then be used to hold the other access control cabling, although it may contain other mains service cables.

Steel conduit generally comes in lengths of 3 m or 4 m and in diameters of 20 mm or 25 mm and should conform to BS 4568.

If the access control system cables are to be carried in conduit adequate room must exist for them to be drawn into position without damage. In real terms this is governed by experience if problems at a later stage are to be avoided.

Allied to steel conduit is the use of screwed fittings as these enable the continuous runs of conduit to follow designated curves and paths and to overcome the necessity for separate runs for each branch, or fittings such as elbows and 'T' pieces must be employed. These are all supplied with their connecting points female threaded to allow the conduit to be screwed onto them, and are available in a huge variety of forms to allow maximum efficiency of any installation. They are all classed as either 'inspection' or 'non-inspection'.

Inspection versions have a small cover that can be removed enabling the withdrawal or addition of new cables.

Non-inspection types are solid, and elbows and 'T' pieces have restrictions in use that are mainly related to the number of cables that can be drawn into the conduit since greater lengths result in higher stresses.

The conduits are joined by screwed couplers or components called running joints, where it is not possible to turn screwed conduit.

Conduit boxes are intended to facilitate the drawing in of wires by allowing access at sharp bends. These are made in a number of patterns so that once the conduit is taken to the termination point a pressed steel conduit box can be employed. These are sunk into the building structure and have knockout sections enabling the cables to be brought into them via brass glands to achieve continuity. These metal boxes, intended for sunk work, should not be surface mounted. A metal conduit box is adopted as the back box for the system spur unit if the mains cable has been run to the control panel or power supply in metal conduit.

Fixing is governed by the method of installing the conduit.

In exposed positions fixing is best performed with spacer bar saddles as these hold the conduit clear of the building material and allow surface water to drain away.

In internal applications use standard saddles or clip or half saddles. Multiple saddles can hold extra conduit runs with a single fixing. Distance saddles may be used on irregular surfaces and in damp environments.

If conduit is run beneath floors it should be fixed to the joists with metal saddles. When going across joists the joists are to be slotted to the minimum size and close to the bearing of the joists.

If steel conduit is being used the electrical continuity is important so the joints and ends must be cleaned down to bare metal prior to the joint being made. The correct bushes and glands are to be used and screwed ends tightly threaded; all nuts and screws are to be tightened effectively and bonding leads applied. If a watertight system is needed the joints should be painted with metallic paint and gaskets fitted under all covers.

The spacing of supports for rigid metal conduit is given in Table 6.1.

Table 6.1 *Spacing supports for rigid metal conduit*

Nominal conduit size (mm)	*Maximum distance between supports*	
	Horizontal (m)	*Vertical (m)*
<16	0.75	1.0
>16 but <25	1.75	2.0
>25 but <40	2.0	2.25
>40	2.25	2.5

Non-metallic conduit

Although steel conduit provides excellent mechanical protection to any cables installed within it, and also gives earth continuity, it does have certain disadvantages. These problems relate to condensation, rusting and corrosion that in the long term can lead to diminished protection and loss of efficient electrical continuity. For this reason non-metallic conduits, which are of rigid or non-rigid forms, can be considered if electrical continuity of the containment is not necessary.

Although non-metallic conduit tubes do not have the resistance to cutting offered by metallic conduit they do have high impact resistance being manufactured from PVC. They are also dimensionally stable and are not susceptible to water condensation.

PVC conduit is available in similar sizes to metallic conduit. The most popular type is rigid conduit conforming to BS 4607. It has plain bored ends and is of standard or light gauge thickness and tends to be coloured white or black.

Rigid standard gauge conduit is best used for surface installations and where mechanical damage is a distinct threat. The normal method of joining and applying fittings is by push fitting. The push fit conduit entry ensures a tight, reliable fit and when used in conjunction with PVC adhesive an even stronger permanent joint can be achieved. If damp conditions are encountered solid rubber gaskets can be added to the assembly. It is not normal to thread PVC conduit because it weakens the wall of the conduit tube.

Light gauge conduit of rigid form is used with plain fittings and cement is applied to the joints as required. This form of conduit should only be used for concealed work, although it can be used for surface work if there is only a slight risk of it being impacted.

Non-rigid or flexible plastic conduit is available in long coiled lengths in the order of 25 m. It is only used for sunk or concealed containment of cables where its appearance is not important. It has great flexibility and enables awkward bends to be negotiated as it can be threaded through holes easily. This non-rigid PVC conduit can be applied over irregular walls and other surfaces without difficulty and can withstand the stresses imposed upon it when floors are awaiting screeds.

Flexible conduit is normally of 20 mm diameter. It is joined by using a small length of the conduit as a sleeve, having first cut a small slot into it to open it out slightly. The conduit tubes to be joined are then pushed together using the small sleeve as a restraint. Following this procedure the connection should be sealed with a compound for strength and watertightness. To this end fixing kits are available with the conduit and contain the sealing compound. A variant on both rigid and flexible conduit is the oval-shaped tube that is advocated if the conduit is to be buried in shallow plaster.

Non-metallic conduits are extremely popular and there is a huge range of fittings and accessories available. The range includes couplings, reducers, bushes, 'T' pieces, bends and elbows together with a vast variety of junction boxes for use in line with the conduit runs. These items can also be used in conjunction with components such as inspection bends, which have detachable covers. These fittings and accessories are intended to make installation as easy as possible, and to allow branch conduit to run off from each other, and inspection and drawing of cables through the conduit system at a multitude of points.

Effects of temperature Although PVC causes little internal condensation and is not a problem, temperature, however, is a consideration. Rigid PVC conduit and fittings are not intended for use when the ambient temperature can fall below –5°C or where the normal working ambient is in excess of 70°C. In cases where the ambient heat is only at these levels for short periods it is possible that PVC may be acceptable, although it is wise to verify the conditions with the manufacturer who may have test data.

In low temperatures the PVC becomes harder and less ductile, and is then more susceptible to damage from impact. The linear coefficient of thermal expansion for PVC is 6×10^{-5} to 8×10^{-5}°C and this equates roughly to an expansion of 12 mm for a 4 m length for a temperature rise of 45°C. For this reason expansion couplers should be used when rigid conduit is installed in straight runs for lengths in excess of 6 m. An exception is where bends are employed as these tend to compensate for the expansion. The saddles which are used as the supports for the conduit can also be set to permit a measure of lateral movement.

The spacing supports for non-metallic conduits are shown in Table 6.2. This relates to the saddles and clips that are fitted and allow longitudinal movement to cater for heat expansion. However, if the ambient

Table 6.2 *Spacing supports for non-metallic conduits*

Normal conduit size (mm)	*Maximum distance between supports*			
	Rigid		*Flexible*	
	Horizontal (m)	*Vertical (m)*	*Horizontal (m)*	*Vertical (m)*
<16	0.75	1.0	0.3	0.5
>16 but <25	1.5	1.75	0.4	0.6
>25 but <40	1.75	2.0	0.6	0.8
>40	2.0	2.0	0.8	1.0

temperature is high or the area is subject to rapid temperature changes the fixing centres are to be suitably reduced.

Fixings should be made at 150 mm from bends, and good aesthetics can be achieved by measuring between supports in long runs and keeping these equidistant.

Trunking and flange trays

The erection of conduit must always be made before the cables are terminated and in position; however, a great advantage of trunking lies in its ability to hold cables that are installed before the containment is made.

PVC trunking is widely used by the access control engineer and can be applied at the last stage of an installation or to an existing application. This is a form of channel manufactured from high impact PVC and featuring a locking or double locking lid that is pushed into position and held within its longitudinal channel ridges.

Trunking is normally supplied in a white finish and is ideal for a range of electrical installations. It is manufactured to comply with BS 4678 and can either be held by fixing its back channel section in place with plugs and screws or by use of a self-adhesive foam strip if the surface to which it is being applied is free from dust and grease.

Outlet boxes to complete the installation are also available to ensure an adequate level of mechanical protection.

Trunking tends to come in lengths of 3 m and is easily cut to size. The most popular sizes encountered are:

Width (mm)	*Depth (mm)*
16	12.5
16	16
20	10
25	12.5
25	16
38	16
38	25
38	38

As with conduit, a variety of accessories can be purchased to make up the installation. These range from couplings that are used to join the lengths of trunking to box adaptors, 'T' pieces, flat angles and internal and external angles to negotiate bends and corners or make up branches. Blank ends are used to terminate runs and to enclose the end of the trunking for strength and aesthetic purposes.

A range of PVC mini-trunking can also be sourced, which is supplied in a dispenser box as a flat coil some 15 m in length. It is easily cut to size, reducing waste and the need for excess joints and couplers. It is fixed to the required surface in a flat form and the sides may then be folded up and the lid clipped into place. Once again, accessories are available to extend its use, including outlet boxes.

In the industrial environment galvanized trunking and fittings may be preferred as they are extremely robust. They are manufactured from precoated galvanized steel with the lid fastened by integral fixing bars that engage the trunking body when the captive screws are rotated through 90°.

A further product worthy of note is the standard flange tray that is used to carry heavy cables. These are similar to trunking but do not have a lid and are found mounted in a horizontal orientation. These often carry the mains supplies within the building and data cables should not be run close to existing galvanized trunking or flange trays carrying this wiring. They may be used to support the mains supply for the access control system.

Dado or bench-type trunking can also be considered to hold data cables. This is installed as skirting and is an effective means of running cables at a low or bench height. This form of containment tends to be manufactured either in PVC or sheet steel with a white epoxy paint finish. Again accessories to complement the trunking can be found.

Aluminium tubing

This is not as popular as conduit but can be used to protect individual cables in some external applications where appearance is important but trunking is not acceptable as a more total enclosure of the wiring is wanted.

Aluminium tubing does not corrode and has a role to play in harsh environments. It is available in 3 m lengths and is easily cut to size. Aluminium tubing is joined by clamp couplings that feature screws and bolts that fix the coupling ends over the tubes being joined. Elbows to negotiate bends and changes of direction are also applied by a clamping technique.

Saddle clamps are used to fix the tubes to the building structure.

Capping/channelling

PVC-sheathed cables running along walls may be buried directly in the plaster of the building but are better protected by placing metal or PVC capping over them. This gives a level of resistance to damage occurring at the plastering stage or just prior to it when construction is still being performed. Essentially it is used for the protection or containment of electrical installation cables when surface wired to brick and block work prior to plastering or rendering.

Generally supplied in PVC in a variety of lengths, it is so flexible that it can be coiled and carried on reels. It is also shatterproof so it does not crack if nailed in position. It takes up the contours of the wall and is easily cut to length. It is manufactured from insulated self-extinguishing material having low smoke, low toxicity and low acid emission properties.

Various widths are available to accommodate different numbers of cable runs. Once the cables are run into position the capping is placed over them and nailed into position at its outer flanges. It also aids the plastering process in that the cables are securely fixed and supported before this process is commenced.

Although metal capping is available it has been largely superseded by its PVC counterpart as the latter type is adequate, certainly in the majority of applications. An advantage of metal capping, however, is its resistance to being drilled at a later stage, by a tradesman unaware of the position of the building cabling, so it does afford better protection than PVC.

Having considered that in some areas cables must of necessity be protected from mechanical damage this does not apply in all installations as the cables may not be accessible or they may be concealed. They do, however, still need to be supported.

Table 6.3 *Spacing of cable supports for PVC insulated cables in accessible positions*

Overall cable diameter (mm)	*Horizontal (mm)*	*Vertical (mm)*
<9	250	400
>9 but <15	300	400
>15 but <20	350	450
>20 but <40	400	550

In this case the cables are to be fixed at intervals as listed in Table 6.3. This table refers to PVC-insulated cables but the spacing of supports can be applied across the range of cables.

Table 6.3 relates to cable supports and this is by means of clips or saddles. PVC clips with a single hole fixing are the most popular, with the internal surface of the clip being formed to suit the cable size and shape. Self-adhesive clips can also be used where it is not possible to drive the nail fixing into position but the surface is otherwise smooth, solid and free from grease.

In the event that the cables are installed in normally inaccessible positions and are resting on a reasonably smooth horizontal surface then no fixing is required. However, fixing is to be applied on vertical runs that are greater than 5 m. At the cable installation stage it is necessary to apply incombustible material to prevent the spread of fire. Tubing should be used to negotiate the wiring around sharp surfaces, including brickwork.

Cables under floors must not be installed so that they can be damaged by contact with the ceiling or floor or their fixings. Wiring passed through drilled holes in joists must be at least 50 mm vertically from the top or bottom and be supported by battens over extended length runs.

Cables passing through structural steelwork must be fitted with suitable bushes to stop abrasion.

Other considerations ask that care be exercised when removing the sheath of the cable and that this be of the correct length to allow cables to be pushed back into position after terminating without too much tension or excess slack. Conductor ends must never be reduced as this leads to a loss of current carrying capacity.

Following on from fixings certain cables must be run underground or overhead to their destination.

Underground cabling

In these instances suitable ducting should be employed, which is sealed with a non-combustible material after the cables have been drawn into position.

Steel wire armoured cable should be specified as this features galvanized mild steel wires embedded in a polymer moulding to form the armouring and this is further covered in a PVC outer sheath.

For increased security of the cabling rot resistant buried cable warning tape or warning brickwork can be applied above the ductwork.

At the entry point to the building the ductwork should be sloped down to the outside with sealing carried out to stop the ingress of water into the premises.

Overhead cabling

Although cables run between buildings across free space can also be held in tubes a more practised method is to use a catenary. Such a technique involves the fixing of a support wire, normally of galvanized steel cable, at each end to a rigid structure and at a height allowing for the passage of persons and vehicles beneath it.

This is used to suspend the access control cables by galvanized clips spaced at some 300 mm but with a loop of slack cable left at each end to cater for movement attributed to changing weather conditions. The cross-sectional area of the support wire is dictated by the cables it must hold with an allowance for the extra weight of any ice forming on it.

Having looked at the subject of the installation of the system cabling it remains to say that once the cables have been run into position they should be identified within the equipment. A good method of carrying this out is by identification tags and this should be followed by the marking up of drawings to help fault finding in the future.

Before we consider the cabling of the mains supply to the control equipment we can make some general observations on the system cabling as these are time honoured and apply across the spectrum of electrical engineering.

- Cables are to follow contours.
- Cabling should not be run closer than 10.5 cm to any fixing point such as the corner of a ceiling to a wall.
- Cables are to be sited away from door frame uprights.
- All wiring should travel in straight lines.
- Never run cables diagonally across walls.
- Apply containments to cable if they are likely to suffer damage.

- Do not use jacketless cables and sleeve the cables in a containment with the same resistance to fire as the building material when passing through a floor or wall material.
- If cables are buried use armoured cables in ductwork.
- Do not pass cables close to steam or hot water pipes.
- Data and communication/signalling cabling should be kept isolated from mains cabling and heavy current apparatus and wiring and only cross it at right angles.

Inspection and testing of the mains supply

All access control systems are permanently connected to the mains supply, although batteries may be used as a secondary supply for standby purposes. There are many precautions to be made prior to making the mains supply connection and there are certain inspections and tests that should be carried out. The Electricity at Work Regulations require that people who conduct these tests are competent under the terms of the Health and Safety at Work Act, so the person who carries out these tests must be able to provide evidence with respect to the required competences.

In the first instance we should look at how the mains AC supply is derived.

Power is generated almost everywhere as alternating current (AC) which means that the current is changing direction continually. In the UK this change of direction occurs 50 times per second and is classed as 50 Hz, the number of cycles per second. The generator that produces this current has three sets of windings with one end of each winding connected to a common or star point called the neutral. The other ends of the windings are brought out to the three wires of the phases or supply cables. For identification these are coloured red, yellow and blue and the currents which are transmitted in each phase have a displacement of 120°.

Power supplies to towns or villages are provided from the power stations via a system of overhead and underground mains and transformers that reduce the voltage in steps from the high transmission voltage to the normal mains voltage at the consumer end. From the substations, low voltage overhead and underground mains are taken to the consumers' supply terminals with the AC power being distributed mainly on three-phase networks. Therefore on the low voltage side of most local transformers will be found four terminals – the red, yellow and blue phases and the neutral. Between each phase and the neutral there is a voltage of 240 V but between the phases the voltage is 415 V. For this reason the secondary output voltage of transformers is given as 415/250 V. The supply authorities are required to maintain the voltage at the consumers' supply terminals within +/−6% of the nominal 240 V.

Most industrial and commercial premises will have a three-phase supply but in domestic premises the supply will only be a single phase unless high loading could be predicted.

The electronic access control installer will appreciate that power transformers are more efficient if the load on each phase is almost equal, so single-phase services are normally connected to alternate phases and three-phase consumers are encouraged to balance their loads over the three phases.

The local boards provide underground cable or overhead line services that terminate at a convenient point within the premises. Overhead services are terminated on a bracket high up on a wall of the property and insulated leads are taken through the wall to the meter position. Underground services tend to be brought through floor level via ducting. The overhead or underground service leads are taken into the mains fuses. From this fused point the supply is taken to the meter and when three-phase supplies are provided, three mains fuses are used with one composite meter.

With underground services, either the lead sheath of the cable or the wire armour is used to provide an earth. A separate wire is normally bound and soldered to the sheath or armour at the terminal position and then taken to an earth connector block. All protective conductors in the property are taken back to this block.

With overhead services an earth block may be provided if protective multiple earthing (PME) is adopted. In other cases an earth electrode in conjunction with a residual current circuit breaker must be used. In some cases a separate overhead earth conductor is provided.

In large blocks of flats or offices the services to each floor are provided by 'rising mains'. In these cases it will be noted that the consumers' meters are located on the individual floors and the rising mains are used to carry the bulk supply up the building. Subservices are teed off at the various floor levels.

In the domestic sector we will encounter only a single-phase 240 V system but in most other instances multiphase supplies will be found. To this end it is vital that the supply is correctly inspected and tested and precautions must be taken to work across one phase only. Equally when remote power supplies or separate mains feeds are being taken to different points of equipment of an access control system it is advisable that they remain on one phase only as the load is not high. This is not vital if single door controllers are used. Ancillary equipment can be an exception to the rule if plugged into standard sockets and this includes VDUs and printers.

The requirements in which we are interested extend to earth continuity, polarity, insulation resistance and earth loop impedance. After completion of any new mains wiring system or addition or alteration the supply should be tested and inspected to confirm that all necessary conditions have been satisfied. However carefully an installation has been completed it is always possible for faults to occur at a later stage by damage being made to cables by nails or other building materials, or by connections being broken or defective apparatus being installed. A full test of the mains connection by the installer should be carried out and the results carefully recorded.

In the UK the procedures for inspection and testing on completion of an installation are covered in part 7 of the IEE regulations, which includes a checklist for visual inspection. This visual test covers the mechanical protection of the cable and housings. This is to check for damage to insulation or for broken or incorrect connections or for breaks or cracks in junction boxes. It is also necessary to ensure that the cable route is satisfactory and the wiring is adequately supported and the correct capacity of wiring has been used. The cable itself must be mechanically and electrically protected against fault conditions.

Having satisfied the visual examination we can then look at the earth continuity, polarity, insulation and earth loop impedance. In practice the requirements of the IEE Wiring Regulations and BS 7671 extend well beyond these tests but the access control system will only make up a small portion of the full electrical system of the premises. For this reason, providing all of the other appropriate tests have been conducted, we can be assured that the continuity of protective conductors and bonding of earth connections for safety have been guaranteed.

Earth continuity

This test is to ensure that there is continuity to the main earthing terminal at each outlet on the circuit. It is performed by disconnecting the supply and testing between the phase or neutral conductors and the protective conductor at each circuit outlet using a hand operator or other portable device. There are no specific values invoked but it is generally done in conjunction with the earth loop impedance test.

Polarity

This test is to confirm correct polarity in that it checks for non-reversal of the phase and neutral conductors and also that the neutral is not fused. There is no specific test type in the regulations but a bell and battery set can be used with a long roving lead taken back to the phase live lead connected to the load side of the meter and with a short lead taken to the apparatus installation point. If the polarity has not been reversed in the wiring run the bell will ring when the roving and short leads are connected to the same conductor.

Insulation resistance

This involves a test for resistance between the pole or the phase terminal and the earth terminal with a DC voltage approximately twice the normal rms value of the working voltage. The insulation resistance to earth between the linked phase and the neutral connections should not be less than 0.5 MΩ. Often an installation must be sectionalized if a faulty circuit is found because the insulation resistance nay be generalized in the system. For this reason it is always advised that the electronic access control system mains supply is a unique cable run back to the consumer unit and therefore not governed by other wiring in the building.

Earth loop impedance

This in practice is the earth fault current loop. A test is done from the phase conductor to the protective earth conductor, which is the path that would be taken by any fault current. This would be done as a test for the full electrical system with the permissible values of ELFI taken from charts in the IEE Wiring Regulations.

It is important to understand the reason for this test otherwise the access control installer may well make connections to an electrical system that of itself is not adequate at source. In addition the access control system relies on efficient earthing for suppression of interference in the system.

The provision of the mains supply becomes an essential part of the installation of the system cabling and in certain circumstances, for example premises in the course of construction, the main contractor will provide the mains supply to a spur for the access control system. If possible the mains source for the entire system should be from one dedicated fused spur. When such a dedicated supply is available it should be identified to stop it being used for an additional purpose. If it is not possible to provide a single spur for the full system and in these cases it is important to use the same phase. An isolation switch should be fitted to the spur feed to enable the entire system to be isolated for servicing and this should be identified.

Usual practice is to connect the equipment power supply's spur unit within arm's reach of the equipment so that isolation is easily performed before any work is carried out. There is no statutory call for the spur(s) to be direct from the consumer unit but it is important to ensure that it is not possible for someone inadvertently to isolate or restore power from another source. The engineer must never work with any electrical equipment live to the mains supply and must power down before modifications are made. This applies when changing circuit boards when any batteries supplying secondary supplies must also be disconnected.

The cabling of the entire network including the signal and data communication should be tested before connection of any equipment. However, modern cables are unlikely to break down and most faults with them occur during the installation process or are caused by other building works on sites under construction.

Certain tests are specialist but those for insulation resistance, in accordance with the manufacturer's data, and continuity and short circuits are readily performed.

Insulation resistance is carried out between the cores in a similar way to that for the mains supply whilst standard test equipment can check for infinitely high resistance with the cores shorted and for shorts with the cores open.

6.2 Siting of access equipment

The cabling installation tends to form the first part of the network installation. Once this has been completed it is possible to install all of the other system components and essential equipment before going on to the terminating of the cables and connection and testing of all the system parts.

There is also a need to understand the access point classifications alongside the siting of the equipment as these govern the security level – these will be detailed in this section.

In practice the manufacturer of the equipment will provide guidance on the location of their goods and the securing of these components to the building surface. However, there are certain factors that can be used to fulfil general rules when considering the siting of all of the components within the survey and design stages.

- For areas classed as high security it is advisable that they be held within a larger zone and be central of the building.
- The whole area controlled by the electronic access control system can be of different security levels but with the areas needing higher degrees of protection located within the perimeter of the lower security zones.
- High security areas should have restricted access and be limited to the minimum numbers of personnel.
- Vulnerable points holding hardware and protective cabling are themselves best grouped together in restricted areas.
- Equipment not associated with the access control system but of itself needing ongoing maintenance should be held outside of the controlled areas.

- Central processor equipment controlling a site is to be held in a particular section or room and provided with additional security measures.
- Sensitive areas are best located at the end of a passageway and not on a through route.
- All equipment and components should be so located as to allow ready access for servicing and maintenance.
- For areas that may be subject to dust or other environmental conditions that can interfere with the satisfactory operation of any electrical equipment only goods with an appropriate resistance to contamination should be considered. Equally the equipment must be able to satisfy any changes of temperature or humidity it may be subjected to.
- Escape routes and emergency exits must not be impeded by equipment. Any doors modified to operate in the access control system should not interfere with emergency procedures.
- All control equipment should be held in a safe and secure area.
- Tamper detection and alarm signalling should be applied to components that are in areas subject to attack or vandalism.
- Locate all token readers suitably close to their control units and unless other conditions dictate at a height of 110 cm above floor level.
- Install door open sensors as close as practical to the opening edge of any door.

NACOSS codes of practice classify the access points by the requirements for successful legitimate access, i.e. the level of security provided. These access points are defined as the position at which access can be controlled by a door, turnstile or other secure barrier.

The installation company is to indicate to the client the classification of the access points that make up the system, related to the level of security provided.

Facilities to control readers from a central point, to record information regarding the access of individual token holders and to monitor the status of the access point where this is required may be incorporated into any class of access control system.

Class I – Common code

At an access point to Class I, access will only be granted following the input of a correct code. This may be numeric, alphabetic, or a combination of both, with a minimum of four digits and/or characters.

The code used shall be one of not less than 1000 differs and is to be protected against unauthorized change and attempts to select the correct code.

A common code is a sequence of characters unique to a particular keypad operated system and allocated to every user of the system.

Class II – Common token
At an access point to Class II, access will only be granted following the presentation of a valid common token to a reader and when the code within the token is recognized by the system.

Each token is to have the same encoded data chosen from a minimum of 10 000 differs. Each code is to be protected against unauthorized change

A common token is one that is unique to a particular access control system or reader so that all user tokens are identical.

Class III – System token
At an access point to Class III, access will only be granted following the presentation of a valid system token to a reader.

The token shall be encoded with a system code of not less than 10 000 differs. The codes are to be protected against unauthorized changes.

The system token is a common token encoded additionally with specific system identification data.

Note 1. Tokens can be added or deleted from the system
Note 2. System tokens should not be acceptable to other systems in the same geographic area unless specifically intended to be so.

Class IV – Unique token
At an access point to Class IV, access will only be granted following the presentation of a valid unique token to the reading device.

The token shall be encoded with a minimum of 10 million differs. The code is to be protected against unauthorized change.

The unique token is one which carries some data allocated uniquely to the user of that token in addition to common data for all of the system users.

Note. Tokens can be added or deleted from the system.

Class V – Unique token and PIN
At an access point to Class V, access will only be granted following the presentation of a valid system token (see Class IV) and the input of a correct PIN of not less than four characters.

6.3 Connection and testing of system components

At this stage, following the siting and securing of the equipment and components, it becomes possible to terminate the cabling and then test the system components.

Cabling

The first step is to consider the normal interface between the equipment and the wiring such as the wiring connectors, terminal blocks or the other methods by which the cables are to be terminated. It must be said that bad connections lead to high resistance and voltage drops.

It is actually at these interfaces that so many problems are caused. Copper, which is the standard material for interconnecting wire, is supplied 'half hard' meaning that it is neither too hard or too soft but is in a condition that despite being incredibly strong is also easy to bend into shape. When bending copper the inner portion becomes compressed whilst the outer is stretched making it slightly 'work hardened'. In consequence the copper loses a certain degree of its flexibility. For this reason there are a number of factors that must be taken into account when using any screw type connections.

Terminal connection blocks

This is the most popular technique.

Wire strippers are to be used to remove the cable insulation to leave only 1 mm once the conductor is inserted into position. If stranded cable is used to avoid loose strands they should be wrapped tightly.

A consideration with this jointing method is that terminal blocks with pinch screws for securing the wire in the block compress the wire over a small concentrated area and make it a little more brittle. Providing that the wire is not disturbed the connection will perform reliably. However, if there is a need to remove it during servicing or fault finding it can fracture when it is reconnected. In practice the thinner the strand of wire the less work hardening when the wire is bent but the lower its mechanical strength and current carrying capacity.

Single strand wires fracture more easily than multiple stranded thinner wires twisted together. Indeed for greater reliability stranded wire can be used in conjunction with terminal blocks that have flexible strips between the wire and screw in preference to the screw end rotating against the copper conductor.

The terminal connection block is the most popular and reliable form of termination within equipment if correctly used, but there is an alternative method.

Clamped joints

This covers the securing of a conductor under a screw head in preference to holding the conductor within a connector block at the end of the screw. As with connection blocks use wire strippers to remove the cable insulation. After stripping, twist the bared wire in the same direction as its natural twist and ensure there are no loose strands if multistrand cable if used. Use a clockwise wrap under screw heads making sure that only some 1 mm of bare wire is exposed beyond the screw head and always support the wire when tightening the screw into position. Ensure that the conductor is firmly gripped but not overtightened or damaged.

With both techniques of using screw holding connections the outer sheath of the conductor should not extend outside of the equipment enclosure and a little slack wire should be allowed to enable the equipment lid to close without stretching or trapping the cable.

There are three other techniques of terminating or jointing that should be known. These are crimping, soldering and splicing.

Crimping

This is usually considered as an alternative to soldering.

Stranded wire crushes down well, so crimp-on connectors are held firmly. Crimping tools used properly with the correct pressure applied will always provide good electrical and mechanical connections. There is a good range of crimping connectors available and the task of applying them is both quick and easy.

Soldered and crimped connections can also be made up in particular junction boxes which can have tamper protection to stop unauthorized access into the cabling.

Soldering

Soldering irons range from small gas powered devices through to miniature units either powered by 12 V batteries or from the mains 240 V supply with ratings from 15 W up to the higher capacity 50 W soldering stations.

The most popular solder is referred to as 60/40, being an alloy of 60% lead and 40% tin. Its melting point is low enough to allow safe soldering of most heat sensitive electronic components and is used with a non-corrosive flux that automatically cleans away oxides formed during the soldering process. It is available in a solder reel pack in different gauges to suit the parts to be soldered. The melting point is in the order of 190°C and the gauges available are generally 18 SWG (1.22 mm) and 22 SWG (0.71 mm).

To carry out the soldering process there are a number of time honoured practices:

- 'Tin' the soldering iron tip by melting a little solder onto it after first cleaning the surface area.

- Ensure that a mechanically strong joint has been obtained by twisting the wires together so that when the solder is applied it will lock and seal the surface.
- Hold the iron to the joint for a short period to preheat the joint and add the solder to the joint. If the solder does not melt, withdraw the soldering iron and apply more heat to the wires.

A good solder joint is shiny and smooth. A bad 'cold' joint is dull and may be rough.

When wires are joined and soldered the solder should flow with a smooth contour to meet the wire at the ends of the joint. If it forms a blob with thick rounded ends the joint is not satisfactory.

An excellent level of mechanical strength can be obtained by a good soldered joint together with good electrical conductivity. The installation can be complemented by the application of sleeving being drawn into position. To this end, heat-shrinkable tubing be applied, which will shrink to some 20–50% of its original diameter after the application of heat by a hot air blower. These sleeves have a high dielectric strength, resist attack by solvents and alkalis and can even provide a moisture-proof seal.

Splicing

This is a joint made by tightly winding the ends of two wires together for a short distance and then mechanically restraining it by an acceptable method such as soldering, crimping or sleeeving.

Although we have discussed the standard techniques of the terminating and jointing of cables there is a further type of cable that must be considered, namely the screened cable used for the transmission of data signals. The screened cable type is identified by an outer sheath or braid of stranded metal wire that encapsulates the cores within the cable.

Screened cable connections

The cable screen or braid runs the total length of the cable and should be twisted to form a cable/pig tail that must be kept to a short length and then terminated close to the entry point of the equipment enclosure. The integrity of the screen must be maintained throughout the system as an unterminated screen will act as an aerial and pick up interference. It is to be terminated close to the cable entry of the enclosure to reduce the occurrence of radiating induced interference over the electronic components held within it.

This screen should be terminated at one end only and at the control point otherwise it can cause earth loops to form between different items of equipment in the system.

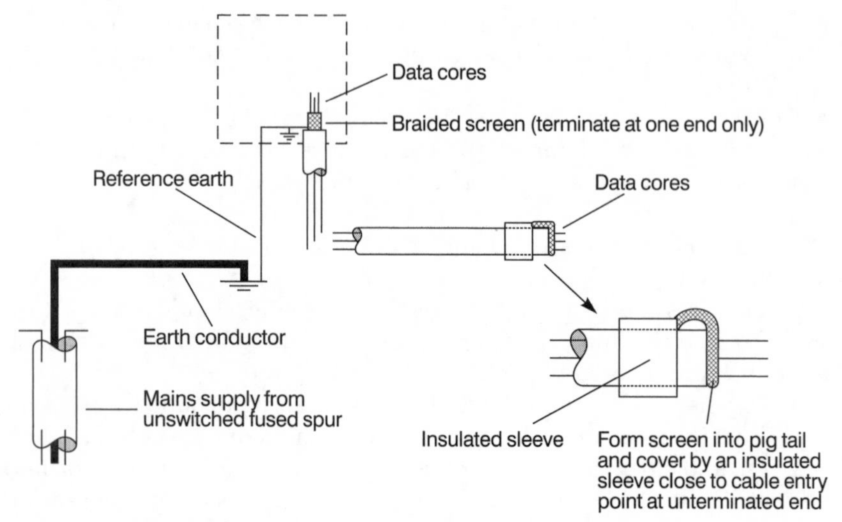

Figure 6.1 *Single point and reference earth screened cable termination*

The screen itself is taken to a reference earth that is the path for induced current to dissipate, although this reference earth is small in relation to the earth protective conductor that is needed to blow a fuse or operate a residual current operated device.

At the unterminated end of the screen it should be laid back along the outer jacket of the cable and then sleeved by an insulating tube.

Figure 6.1 shows the relationship of the reference earth and single point screened cable termination technique. Correct earthing is needed to ensure that all exposed metal parts are at 0 volts potential and the requirements of the IEE Wiring Regulations are upheld.

Cables and interconnections must be installed for reliability and although a fault in a connection may be visible, problems with electrical interference and failure to suppress this at the component connection stage form a more complex subject.

Electrical interference is liable to be environmental and can be conducted, induced or radiated. Although we have legislation covering electromagnetic compatibility (EMC) and the restriction of it to interference beyond prescribed limits there can still be site problems encountered with it. Equipment is clearly designed with inherent suppression but there are further precautions that can be made to combat its effects.

Conducted interference is in the main generated by high speed electronic switching of power equipment that is connected to the mains supply. This mains borne interference can then be induced into any electronic equipment and creates erratic switching of its electronic functions. This tends to be suppressed at its source and not allowed to enter the mains supply, with suppression also built into the electronic control equipment at the point of manufacture.

Induced interference is generally continuous and interferes with the wanted signalling of the electronic equipment. It does also originate from the mains power supply, tending to be induced into signal and data cabling from adjacent high power mains transmission wiring.

Radiated interference comes from lightning or electrical storms and radio transmitters. It is also a consequence of discharges such as that generated from fluorescent lighting. It is best dealt with by the addition of low inductance capacitors connected close to the components most susceptible to the interference and a ground point.

It was mentioned earlier about single point earthing and the connection of the braided screen in screened cable to avoid the malfunctioning of equipment from interfering signals. A consideration of the earthing is that the screen be sheathed so that it cannot make metallic contact at a further point along the line. The screen is usually earthed at the power supply/ control end only. The earth conductor at that point should in effect be a bond between all exposed metal in the premises and the final earthed position to ensure that an electrical path exists that will operate current leakage protection in the event of a short circuit of the mains supply. The reference earth is different in that it is only a path for induced current to flow from the screened braid of a shielded cable, so is small in size compared to the main earth conductor.

Further reference should be made to Section 4.6 as this deals with cable protection and filtering.

Another form of suppression to overcome is that of spikes created by an electric field collapsing when DC loads are switched by devices such as locks. If these are not suppressed the spikes can enter the supply cabling causing corruption and damage to outputs and other sensitive switching contacts.

When power is removed from an inductive device such as an electric lock a large voltage or back EMF appears across it and this can be in the order of several kilovolts. This can conduct along cables to cause malfunction of various components so the spike should be suppressed locally at source.

Figure 6.2 shows the normal procedure to suppress the discharge using a signal diode in parallel.

The cathode is connected to the lock positive connection close to its lock terminals so that when the power is removed the back EMF is conducted in reverse through the diode and the energy becomes dissipated in the lock itself.

At Figure 6.3 a method of series/parallel connection is shown, which ensures that damage is avoided to the diodes if the supply voltage is connected in reverse.

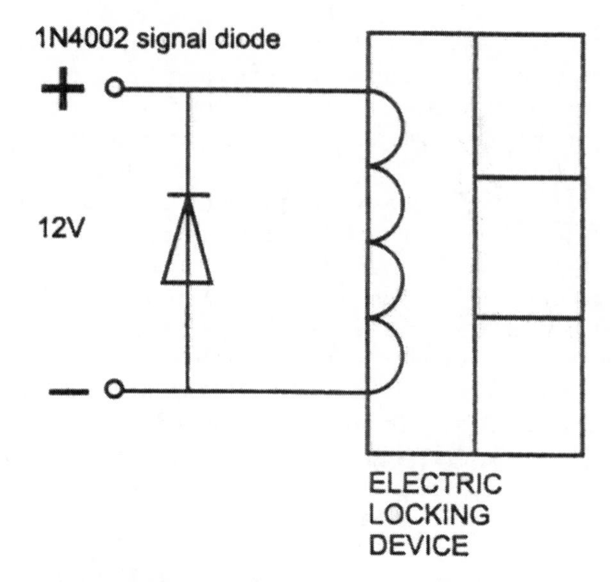

Figure 6.2 *Signal diode in parallel*

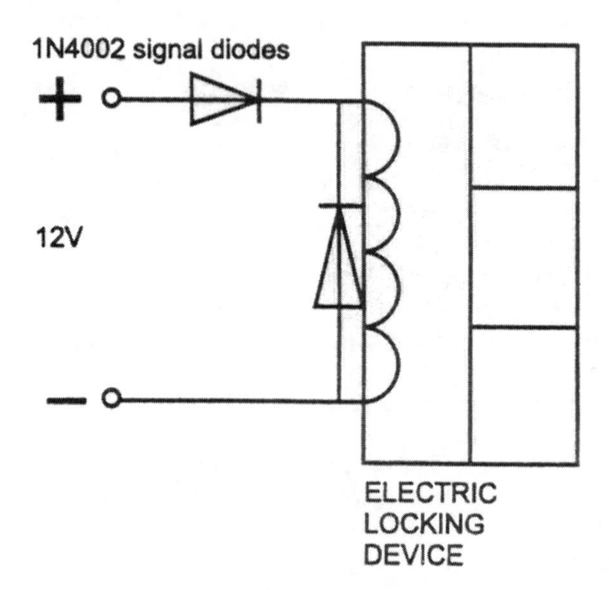

Figure 6.3 *Series/parallel connection*

An option to this is to use transient suppressors as these may be connected either way around and lower their resistance once a transient is detected.

If AC locks are being used in preference to DC versions metal oxide varistors can be used in place of diodes.

At this point we can assess the cabling that has been terminated at the equipment and overview the main considerations that apply across the range of systems.

- The data provided by the manufacturer for the equipment is committed to recommend the cable type and connection details.
- The data or signalling and power transmission capabilities of a cable over an extended distance deteriorate.
- Heavy objects impacted against cables cause damage to it and a reduction in its performance level.
- Cable must not be subject to tight bends. This is even more apparent in data transmission and coaxial cables.
- Random breakage can occur with terminations if terminated a number of times.
- The function of one cable and its signals can affect other signals in different cables.
- Electromagnetic interference (EMI) or strong electromagnetic signals interfere with digital signals and must be suppressed.
- Communications wiring and power wiring must be segregated.
- Use shielding and single point earthing to suppress interference on data and communication signals.
- Only cross power cables and data signalling cables at right angles.
- Do not use the same holes or runs for different cabling.
- Assess the environmental conditions of an installation to determine the correct cabling and protection techniques.
- Use repeaters to boost data signals and remote power supplies to avoid voltage drops.
- Surge protectors may not be adequate to protect microprocessors and an uninterruptible power supply may be necessary.
- Terminate all connections using the appropriate technique and without applying undue strain or force.
- Ensure that cables within a containment are not heavily bunched and are not drawn into position with excess strain.

Equipment

It becomes possible to test the equipment once all of the cabling has been terminated and proved satisfactory.

All equipment should be confirmed as able to withstand the following air temperatures:

Internally sited equipment 0° to 40°C
Externally sited equipment –20° to 50°C

Systems that have been purchased as a kit form of single source tend to have a schedule of tests that can be performed at the commissioning stage because the equipment being utilized is clearly defined. However, with the system that makes use of components from different suppliers the test of the equipment and the commissioning of it will differ enormously between sites and installations. For this reason all the system components should be checked individually and in their own right before testing them integral with the other components as a whole. To this end all of the components may come complete with their own installation and data sheets to aid their test, commissioning and fault finding. The test schedule for the system can be broken down into the following areas:

- Perimeter protection and hardware.
- Tokens and readers.
- Controllers.
- Cables.
- Power supplies.
- Signalling equipment.
- Ancillary equipment.
- Communication equipment and software.
- Power circuits.

Although it is possible to overview a general test schedule for all these areas this cannot be exhaustive because of the diverse nature of the systems and the many different roles they employ. However, there are tests that can be conducted to cover the essential rules, although certain systems may need further specifics carried out.

The electronic access control engineer will very often approach different systems in slightly different ways because of the components that may be used. As an example there will be intercom systems to provide internal communications only in order to provide low security for homes. In other instances access control will be installed for businesses with service entrances or delivery doors but these will be audio/video systems needing to limit access and identification before allowing admittance at the service entrance. Included in this area can be banks, retail stores, pharmacies, ticket offices, hotels and fast food restaurants. Public and private buildings often need audio/video intercoms for people to call for assistance. Included in this sector are wheelchair bound people and the older or infirm person. These intercoms can also provide security by limiting access to areas in dormitories, church rectories, schools, nursing homes and prisons. Indeed there are fully fledged access systems that are effectively in kit form, which are intended for single and double door applications in domestic settings, surgeries and small industrial and commercial premises. These can extend to support a number of proximity readers to include anti-passback and control up to 2000 tags. These kits can have extra features such as duress codes, door release timers, door alarms, door contact shunts, timed zones, timed event memories and include outputs to printers.

It will be appreciated that with such kits the testing of the equipment and system can be relatively easily defined; however, we should consider the checks of more diverse systems although these will be treated as areas in their own right.

Perimeter protection hardware

This area covers the locks and hardware such as closers, sensors and push-to-egress buttons.

The lock that protects the perimeter doors and barriers is a critical component. Check for:

- Correct alignment.
- Correct operation in accordance with the specification.
- Ensure that as the lock is energized and de-energized it performs as specified for emergency purposes.
- The locks failing in the correct fashion as fail locked or fail unlocked.
- Manual overrides functioning smoothly and overcoming any electrical malfunction.
- Ensure that all sensors give the correct response to the door position.
- Check the operation of door closers to ensure they pull the door closed with the correct force.
- Push-to-egress buttons should be verified for operation.
- Perform a test to prove that timers generate an alarm activation when doors are open for longer than their preset period.
- Prove that the door lock remains energized for the timed period.

Tokens and readers

This area confirms the verification of the token, credential or physical behaviour characteristic is performed correctly.

- The token reader must reliably read the credential introduced to it.
- Different tokens should be tried.

- Invalid transactions should be attempted.
- Authorized tokens should generate the correct response and output at the display.
- Authorized tokens should be correctly displayed at any interconnected printer or VDU.
- Voided and forged tokens should be introduced to ensure the correct output is generated.
- Authorized tokens are to be signalled to and recognized by the central controller.

In assessing this we can confirm that the reader itself must provide the following features:

- An indication for access granted.
- Variable time available for access to be made.
- Detection of physical tampering and, for readers fitted externally, protection against malicious damage.
- Response within two seconds of the valid completion of the necessary entry procedure and relocking of an access point if it is not then used within a predetermined time.
- Readers shall be securely mounted in a convenient position for the user adjacent to the access point, but proximity readers may be sited at any point where successful activation will occur.

Controllers

The controller is the heart of the access control system being the processing unit that monitors and then controls the reader and tokens presented to it, although it may at times be integrated with the reader.

Single door controllers have a more limited capability than local controllers used in multi-door distribution systems. In order to carry out the checks of the controllers it is necessary to load the system software and carry out the programming, otherwise the components are not recognized as being in the system.

In practice all of the system functions are governed by the response of the controller. The controller can nevertheless be checked as a standalone device to ensure it satisfies all of the required parameters. In brief:

- Perform the tests as for tokens and readers to ensure correct response and the display of programmed messages.
- On activation of the reader the door hardware signalling should be enabled.
- Inputs and outputs to ancillary equipment should respond as programmed.

- Off-line operation should conform to the specification.
- In off-line checks should establish if tokens can still be read and doors unlocked.
- Memory storage facilities and memory buffers should be verified with the controller off-line.
- Remote signalling should be generated as required according to the programmed parameters.

In checking the controller ensure that the manufacturer's specified environmental conditions have been satisfied in respect of:

- Temperature.
- Humidity.
- Dust and other air contamination.
- Vibration.
- Electromagnetic interference.

Cables

These form the interface between all the system components and will be of a different form depending on the signals that they carry. These signals may be data or communication, alarm or power.

- A visual inspection should be carried out to ensure that cables comply with the specification.
- Ensure that no joints are made outside of junction boxes and that unapproved connection techniques are not used.
- Check for damage to the cores of the wiring and confirm that there is no missing insulation or that it is stripped back too far.
- No points in the wiring are to be stressed.
- Prove the consistency of the colour codes.
- Ensure that the segregation of cabling from other cabling in the building is correct.
- Check for suppression being applied.
- Check cables within containments and that conduit is grounded.
- Verify the wiring routes are to the plans and follow the claimed routes.
- Ensure that the ambients of temperature that the cable is routed through cannot interfere with the performance of the wiring. This should not generally be in excess of 30°C.

Power supplies

Failure of the power supply or the cabling to it can cause a total closedown of the system unless standby batteries are provided. Where this continued operation is essential during mains failure the standby should have the necessary capacity to support the system for not less than the minimum period required by the customer.

The power supply is essentially a transformer with rectifier filter circuits and perhaps also a regulated DC supply with a charging circuit. In effect:

- The output voltage and current for the high and low charge voltage needs to be checked.
- Separate battery outputs should be measured.
- Batteries should be proved as fully charged.
- The battery voltage should be recorded with the mains supply disconnected at source.
- The mains supply to the power supply should be correctly fused and be visually and electrically tested.
- Supplies to the access control system should be proved to be identified at their source.
- The power supply should have the efficiency of the earthing confirmed.
- Ensure that the transformer is incapable of creating a hazard for materials or cables within its vicinity.
- The total current demand of the system at its most extreme condition should be subtracted from the power supply output current as any extra requirement must be supported by batteries.
- Ensure that the total current draw from all power supplies is within tolerance and is detailed on the system record.
- Confirm that standby power supplies operate in the event of mains failure and any required warning of the disconnection is generated.
- If the system is driven by a UPS these must be as specified and proved to be of sufficient capacity to cope with prolonged mains isolation.
- Ensure that the UPS or power supply is in a location where maintenance can be easily carried out and that they are in a ventilated area.
- The capacity of any power supply must be selected to meet the largest load likely to be placed upon it under normal operational conditions. This is not to exceed 50 V.
- If safety and security considerations do not require continued operation of a system during a mains supply failure, the public mains via a safety isolating transformer may be the sole supply for the system.
- The power supply should be checked for location in a position secure from tampering
- For systems with fail unlocked perimeter protection devices additional security must be incorporated for the power supply unit.

Signalling equipment

This part of the equipment testing involves us with the signalling that may be local within the protected areas or to a remote monitoring point or central station. The checks are performed in conjunction with those of the controller.

- Local signalling should prove that any warning device or visual monitoring equipment receives the correct response in accordance with the transmission of a signal from the access control system. This can involve sounders, buzzers and any other display units or slave screens and printers or VDUs.
- Any other security or building system or service integrated with the access control system should be verified as receiving an appropriate transmission.
- Door call units used with intercoms should be tested for audible and visual receipt at all appropriate points.
- Automatic dialling equipment is to be checked for secure positioning in a protected area.
- A check should be made with the remote monitoring point or central station that the message that is to be generated is received.

Ancillary equipment

Although certain ancillary equipment may have been verified in the tests for the other areas, this category includes mechanical and electrical sensors, booster power supplies and repeaters plus devices such as printers and VDUs.

- Mechanical and electrical sensors may be used to check that zones are clear before barriers and doors are closed and these are to be confirmed as providing the correct coverage.
- Booster power supplies and any repeater equipment should be verified alongside their standard counterparts and in the same fashion but also be recorded against the ancillary equipment area.
- Printers and VDUs are to be tested for response with the system and the display of the parameters.

Communication equipment and software

This can form the final part of the equipment test schedule before the readings for the power circuits are logged.

- All data must be checked for correct entry.
- All alarms must be correctly displayed.

- All access levels with the times of access allowed must be verified.
- Operator levels are to be defined.
- Events must be shown exactly as they occur and as specified.
- All automatic systems feature as specified.

Power circuits

Following the tests made on the equipment and on all of the components that make up the electronic access control system, in addition to ensuring that any standby supplies operate in the event of mains disconnection, there remains a need to record the voltages applied at the supply terminals and also at the distant loads.

Manufacturers' installation sheets will contain data on the current consumption of powered devices or the system loads so the engineer can therefore both measure and calculate the voltage at these loads.

In large installations a voltage drop can occur at distant loads and this must also take into account the need for an adequate secondary supply.

The electrical resistance R of a circuit element is the property it has of impeding the flow of electrical current and is defined by the equation $R = V/I$, where V is the potential difference across an element when I is the current through it. The unit of resistance is the ohm (Ω) when the potential difference is in volts and the current in amperes.

Cable manufacturers quote cable data and as an example we can consider a cable with a current rating of 1 A and resistance of 8.2 Ω per 100 m. For a two-wire loop the total resistance is 16.4 Ω. In practice this resistance has little effect on a load that draws little current.

For a load of 20 mA at 12 V the voltage drop is $20 \times 16.4 = 0.328$ V so this would create no problem for this device requiring a 12 V supply at the load if the terminal supply was for instance 13.7 V as the voltage at the load in theory would be 13.372 V. However, bad connections and strands of cable lost in terminating would increase the voltage drop because of the increased resistance. It is important that practical measurements of the values are taken with the calculations used as a guide.

It is necessary that the standby supply is also able to power critical devices in the event of mains disconnection so the voltages in this case should be both calculated and measured.

Reference should be made to Section 5.2 which further considers voltage drop and prediction of current in a circuit.

Having taken some regard of the means of connecting and testing the components that make up an electronic access control system we now come to programming and configuration. But before doing so we can perhaps, as an example, consider a given manufacturer's magnetic stripe swipe card system. This is promoted as a compact standalone system. In practice standalone systems make up the greatest volume of installations. Compact implies that the reader has all of the decision making intelligence within the reader housing so these systems are not difficult to install. However, because the intelligence and lock switching wires are on the outside of the secure area access can be gained by knowledgeable tampering with the system. For this reason the product in our example has been addressed differently by the manufacturer and has a separate card reader and switch control unit. This as we know enables the control electronics and lock wires to remain inside the secure area, thus avoiding tamper problems. An objective of the product nevertheless was that installation must be easy.

The switch control unit is supplied as a PCB that can be mounted inside a standard 12 V boxed power supply that most security installers carry as a stock item. The reader and lock are wired back to the control unit. This product actually has a capacity for 10 000 cards and coloured access zones for multi-door applications. Two readers can be connected for read in and read out purposes. The system can be extended to include exit buttons, time clocks, entry phones and keypads for card plus code entry. Volt-free relay contact can switch most locks and barriers direct.

User cards are issued in wallets with paper shadow cards that are used both as a user record and for voiding the user cards. All system settings are carried out using function cards which are swiped through the readers.

A pre-connected tail of shielded multicore cable is provided as a kit to connect the reader to the switch control unit. When making a joint in this data cable it is important to ensure that the shield remains intact and connected throughout its entire length. The shield should be connected to earth at one end only. This will be at the power supply main earth terminal.

The switch control unit PCB has the card reader terminals along one edge and the power supply voltage and relay outputs along the opposite edge. The PCB operates at 12 VDC and the relay outputs are clean changeover so they can be used for either fail locked or fail unlocked devices. A further relay can be made to toggle on and off by swiping a specific programming card through its corresponding reader so it can cause a latch to select a function such as arming an intruder alarm panel. Push-to-egress buttons of any number can be connected with an external time clock to control access to doors during predetermined time frames. A diode is supplied for connection to suppress back emf from the electric lock. LEDs of four colours are displayed on the reader head to give diagnostic and access level information to the users. The reader head only accepts cards swiped in one direction but the reader face has an illuminated arrow to show the direction in which this must be done.

Essentially this is an intelligent system which allows the use of an unlimited number of standalone reader heads with three levels of authorized access where each level of access is defined by the colour of the LED and the user's card. A user can only gain entry when the colour of his or her card corresponds with the colour of the LED or LEDs on display. Diagnostic LED displays can also show a bad card swipe or that the door release is being energized.

The system programming is performed by swiping a particular function card through the reader. As an example, to select the electric lock to be energized for 12 seconds the user swipes the lock release programming card through the reader once, waits for 12 seconds then swipes through a second time. To remove a card from the system, the user swipes the corresponding paper shadow card through the reader.

The programming of this system is somewhat unique to this equipment but the means of configuring the components are standard practice.

In the following section we look at other means of programming that will be readily encountered.

6.4 System programming and configuration

The actual techniques of programming and configuring the information and setting the adjustable values of components that form any system will be governed by the particular equipment and installation. These techniques in some cases may need to be performed at a number of points whereas with other systems all the data may be inputted at one location. For some general parameters potentiometers are used and for some addressing procedures DIL switches are adopted to generate a binary address code. A unique binary address can be formed by a four-way DIL switch such as shown in Figure 6.4. With the switch in the off position, it has a value of zero.

With the switch at the on position it has a value of 1 2 4 8, reading from right to left.

DIL switches are common for addressing by binary code but other processes of simple configuring can include the cutting out of links or the application or removal of jumpers across pins.

Although the initial testing of the system can be performed by considering all of the components as separate areas it is not possible to do a full system test without carrying out all of the programming so that all of the other system parts are in contact with each other as need be.

Displays can be by LED indicators for basic settings and acknowledgements. For more comprehensive programming, lines of LCD text by means of a keypad are often used to scroll through the menu and programming options. These LCD displays incorporate simple questions

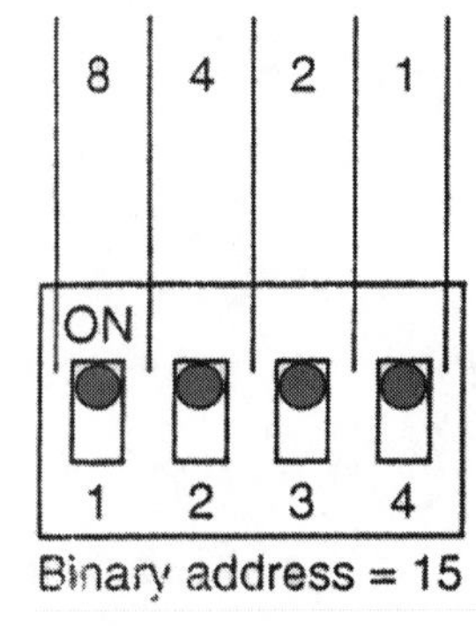

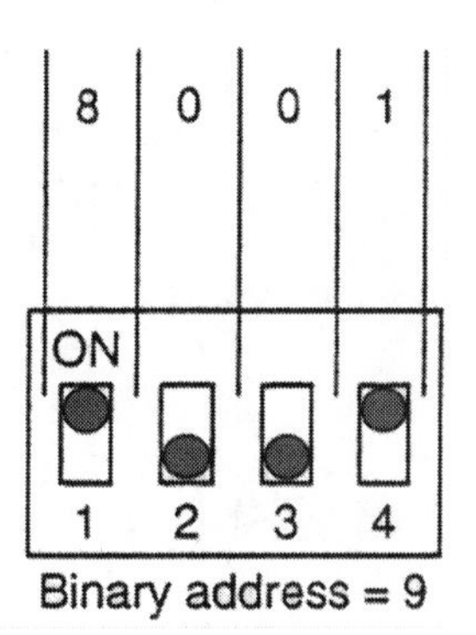

Figure 6.4 *Binary address 4 bank*

so that only minimal reference to installation instructions is needed. No and Yes are often used to scroll up and down the keys allowing the engineer to navigate through the options. If transaction reporting with a recorded event log is to be included in the system a PC or serial printer can be added. Indeed there are benefits for security installers and the end user in linking their access control system to their computer network with the computer being the interface between the employee and the rest of the organization.

The PC is now widely accepted in the working environment and most organizations know that it is an essential component in controlling many plant elements. Indeed on the market there are now specific access control and security management software programs that can control multiple users and doors on a number of local and remote sites. These packages can interface with powerful Windows-based programs and a range of readers to offer full multi-user, multi-masking capabilities with a capacity to provide full monitoring of remote sites through established IT media. This media can include direct connection by fibre optic links, line modem or private line and in certain instances by microwave link.

In general terms we can say that these packages require minimum computer hardware requirements with the system commands accessed through a range of straightforward iconized menus for point and click operation.

An example of 'door' icon accesses door programming is shown in Figure 6.5.

Comprehensive information on-screen can also be obtained with alarm inputs defined on maps. When an alarm is activated, the computer gives an audible signal and the particular map is brought up on the screen. If a suitable sound-card is available, a preset handling message may also be 'voiced'. An example is shown in Figure 6.6.

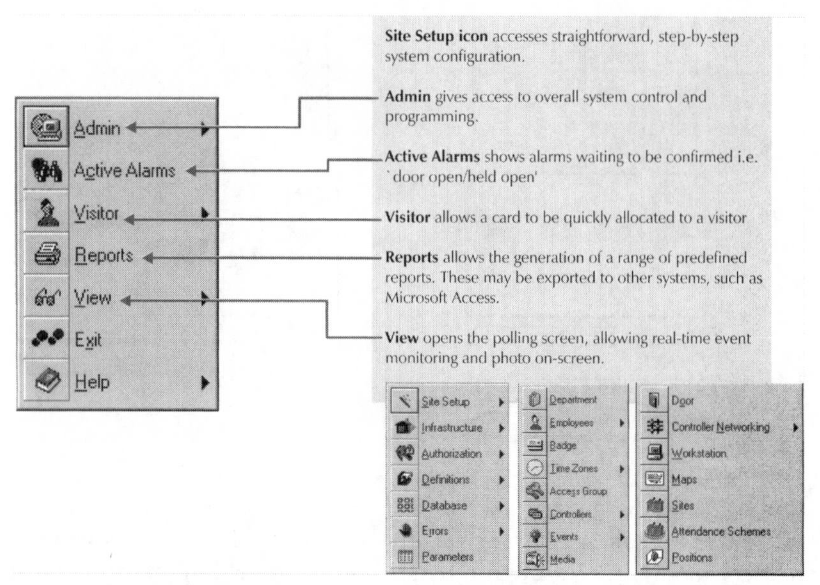

Figure 6.5 *'Door' access icon system management*

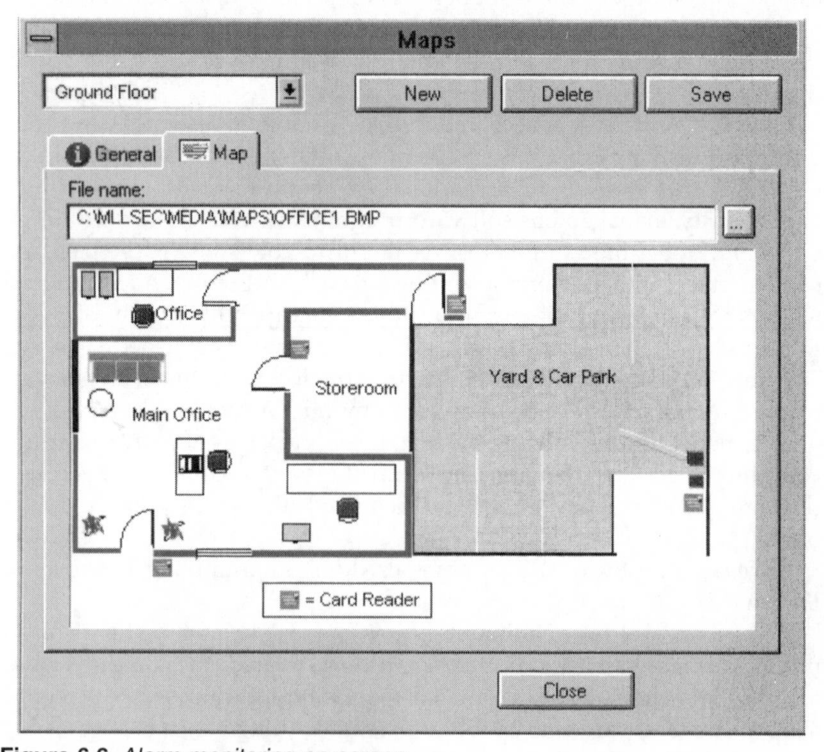

Figure 6.6 *Alarm monitoring on-screen*

Figure 6.7 *Portrait on-screen*

A portrait on-screen may be used to show ID photographs of users. These may appear on-screen together with user information when a card is presented. Figure 6.7 illustrates how card numbers can be depicted with entrants' names and the card status.

The Windows registry is a large and powerful system file but corruption can occur when installing or uninstalling programs so it is advisable to back these up periodically. Floppy disk drives are engineered to accept disks inserted into them and to store the back-up data. These floppies are extremely robust and reliable and are transportable. However, the hard disk in essence is permanently fixed within the mainframe and not normally removed and stores the core programs under Windows or other operating systems such as DOS.

Commands used to perform functions differ under the different operating systems and alternate manufacturers have their own methods of system programming and configuring the system against specification. The electronic access control engineer's responsibility lies in understanding the options of programs available and the different parameters that may need to be programmed into any security system rather than being totally in command of all the available computer concepts.

Name	Department	PIN	Tel no.	Access code	Time code	Card number

Figure 6.8 *Token holders*

We can stress that personal computers are becoming increasingly used as the reporting unit for the larger electronic access control system but are also extending their use into mid-range systems, so basic knowledge is needed.

At this stage we can make some observations on the information that may need to be programmed into any system whether this is inputted into a PC or into a central controller. It can now be appreciated that the techniques associated with the configuring of the data will vary from system to system.

Data must be collected to determine all those who are to use the system and the manner in which the system is to be run. The system programming will take into account such activities as door timers and alarm sensors plus the entry of inputs and outputs. Readers must be configured with automatic operations and there is also a need to enter information on alarm messages and maps.

The programs for those persons who use the system can be covered by listings of token holders, areas accessed, time schedules and management.

Token holders

Input all personnel who are to hold a token. Include token numbers, names, job description and department. Use a planning sheet for easy reference using Figure 6.8 as an example for the headings.

The access codes and time codes should be a list of areas in code form to which the token holders have access and the times at which this access is authorized.

Account must also be made for maintenance staff, visitors, delivery personnel and cleaners, etc.

Areas accessed

Input data for the authorized personnel as to the areas that they can be enabled to enter. This can be a list of the access codes with a cross-reference against detailed descriptions of rooms. Allied to this it is possible to input the doors through which access is granted for the different access codes together with the staff who can then claim authorized passage through those doors.

Time schedules

This involves the inputting of data controlling personnel movement by time rather than by area.

The attendance time of the various staff jobs should be recorded. This can then be related to the staff names and job descriptions that have been inputted. Start and finish times are to be recorded together with days of the week. Allowance should be made for staff who may wish to enter the premises somewhat earlier than the official start time. Overtime, weekend working and holiday entitlements should be separately catered for.

Management

This input refers to the managers plus security and reception personnel. These people can be covered for high level functions such as the ongoing programming of token data with the validating and voiding of credentials plus all the other system parameters. The management are therefore allocated a high security status. Other staff will be inputted at a lower level and only empowered with lesser programming authority to embrace routine activities only.

At various stages in our research we noted some of the programmable parameters of controllers and how these become more progressive as we move from single door controllers through multi-door system networks. We have also noted that off-line denotes an access system not controlled by a computer and on-line refers to a system controlled by a computer to form a network of door controllers.

Standalone systems we may recall have no intelligent links to other control units as they only control a single access point but may provide connections for a few points. These may or may not have event recording. Network systems, however, are systems with access controllers linked together by data cable for the exchange of information between units. This provides easier configuration and better management information for larger and more complex applications. All access points on the system can be set up and managed from one location. In this case the controllers on networked systems record events. These systems allow for a PC connection to enable control and reporting to be achieved from a dedicated program.

This event recording is the memory and includes the log of access point details, date, time and user ID for each transaction when access is authorized. Depending on the system many other events can be recorded but can include access denials and alarm activations. On the occasions when events are recorded the controller must be capable of sending data to a PC or printer for report provision.

Basic programming must take into account time zones which are the time windows when access can be granted, access zones which are the controlled areas and access authority or levels covering identity of the level extent of which users can be allowed into areas of control.

There must also be automatic functions of locking and unlocking of hardware perimeter devices with monitoring for doors open, closed or forced plus door open time allowing the door to remain open for a prescribed period. Allied to this is anti-passback, stopping a user from entering via an access point unless they have already been identified as 'out'. This is to be used in conjunction with an exit reader, or over a programmable time.

There may also be a need for card or door trace history with invalid card or wrong PIN attempt.

It remains to say that the functions and parameters will vary to a great extent between systems and with the more complex network it may be necessary to visit the site later when it is in active use. It can then be overviewed in operation as the management will also have had an opportunity to evaluate its effectiveness.

In this chapter regarding the installation and commissioning of an electronic access control system we have come to appreciate how the security installer must become increasingly aware of the role of the PC and software and the future role these must play. It raises two important subjects: firstly the physical protection of the equipment with intruder alarm signalling, and secondly the backing up of data.

Computer theft is one of the greatest threats to any organization and can only increase. Hardware prices are not changing at the present point in time but with the rise in chip values the central processing units are a prime target. Protection devices can be in the form of entrapments which surround the computer and are securely bolted to the floor or desk. In addition the equipment is easily protected by an intruder system which can have both local audible and remote signalling with perhaps CCTV surveillance monitoring.

The backing up of data on a regular basis enables the rebuilding of a system if there is an irrecoverable crash. The worst scenario is a system crash with all attempts to revive it failing. Windows and the data on the hard disk have been corrupted and the only solution is to format the hard disk and start again by re-installing Windows and all applications. If this ever happened a huge volume of work could be lost and it could take a considerable amount of time to get the system restored to the condition to which it was in prior to the crash. With integrated security, building management and a complex access control system this would have devastating consequences.

There are various backup schemes available that can help to minimize or even eliminate the damage caused by a fatal crash. Making backups is often seen as a nuisance but a few minutes backing up can save files that have taken days or even weeks to create.

Essentially there are two basic backup and recovery strategies:

- Duplicating the entire contents of the hard disk.
- Making copies of important user and access files and re-installing the applications from the original program disks.

In both of these cases it remains important to make backups at frequent intervals to retain the files as current and up to date. This process can be scheduled to take place automatically and to save time it can be performed incrementally so that only the changes to the original backups are recorded.

These backups or copies of backups should be stored in a secure area and at a separate location so that they could even survive as a result of a fire damaging other equipment. In addition these backup systems and the access control data that they hold should be randomly tested for correct operation. This can be performed at the time of the maintenance of the system.

The important thing to remember is that the vital files to back up are those that cannot be replaced, which in essence covers anything written or created on the PC.

Duplicating the entire contents of the hard disk is in practice not easy. It requires a reliable disk- or tape-based recording system with sufficient capacity to store possibly several gigabytes and making the initial backup would be time consuming although subsequent updates mould be much quicker.

Re-installing an entire system to a freshly formatted disk is difficult since without Windows or the original backup program on the drive there is no easy way of transferring the data from the disk or tape and the backup could possibly contain the source of the original crash. This source could be in the form of a bug or virus that could create the problem crash once again. Equally the programmer would be re-installing all of the redundant files and parameters that perhaps were slowing down the previous installation data.

The second recovery strategy is time-consuming but depends on the number of programs and the amount of data that must be restored. The advantage is that the PC is effectively commissioned from new and there is no need to purchase any new hardware or software.

Windows 95 and 98 both have a utility called Backup for archiving important files to floppy disk or a suitable tape or disc-based recording system selected through the Tools tab. A limitation is that Windows must be on the hard disk to run the program so it cannot be used to restore a system to a freshly formatted hard disk.

There is a good range of backup software on the market combining advanced backup and compression utility with crash recovery tools that may save the need to format the disk at the outset.

The access control installer and programmer must determine what is to be backed up. We can say that this must include up-to-date copies of the systems critical files and the Windows Registry. However, they cease to be of use for recovery purposes once the hard disk has been formatted since when Windows is re-installed it creates a new set of system files. Websites should also be included in the backup together with internet and E-mail accounts.

Graphics form a separate subject with imaging devices such as scanners and digital cameras so all these image files must not be overlooked. File compression can be used if there are a lot of imaging devices being employed. These may be gathered together with any other data downloaded from the internet and held in a folder program with the other backups.

Before we conclude the subject of programming and configuration the installer may also want to consider selecting a video badging system – this is Windows based and allows the user to produce durable credit card quality ID cards which are compatible with other security systems. These cards can be issued to staff and visitors through a user definable database managed directly from a PC and can be printed onto current access control cards.

The person's image is grabbed by a camcorder, CCD or digital camera and then stored on the hard disk so that the re-issue of cards is an easy process. Figure 6.9 shows the process.

This badging system can also be extended to vehicle identification for vehicle registration and make/model. The tag or semi-passive transponder in the vehicle windscreen can be read by an RF base station for automated entry and can also hold the vehicle image data for increased security as shown in Figure 6.10.

6.5 System handover

At this point the installer will be at the stage of handing over the electronic access control system to the client. In addition to presenting all of the documents in a professional manner the customer must be briefed on the need for servicing and maintenance.

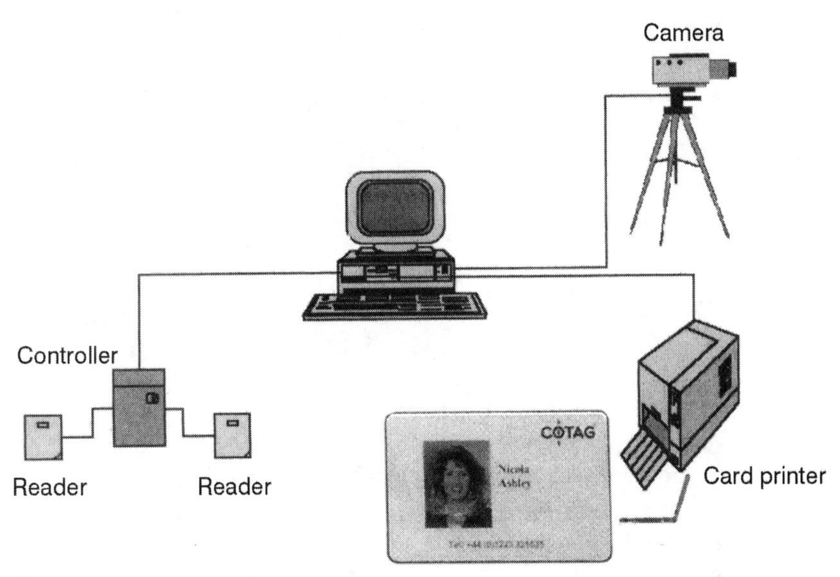

Figure 6.9 *Video badging process*

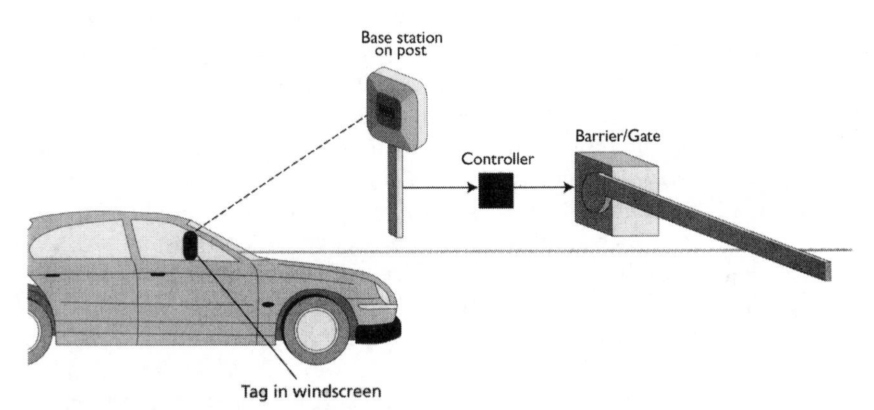

Figure 6.10 *Vehicle identification*

The inspection, testing, commissioning and programming of the system are to be exhaustive. Equally the handover must be comprehensive to heighten client satisfaction. Nevertheless the client must be instructed and advised that they must accept a measure of responsibility for the system to be used in the correct manner. To this end the customer must be furnished with all documents related to essential components, drawings and a record for the installation. These documents must have all necessary operating manuals and instructions with diagrams of the site and perimeter protection. Cable runs can be detailed alongside the equipment location points.

A full demonstration of the operation of the system must follow and any duties that must be performed by the user which are necessary for the efficient running of the system.

The installer should:

- Ensure that the system is exactly as that detailed and originally specified.
- Prove that the workmanship is to the required level and that the system operates correctly. A record should have been established for its full test and commissioning.
- Confirm the effectiveness of the system, paying particular attention to those factors that enhance reliability.
- Accept a measure of the responsibility for the simplicity of use of the system.
- Furnish the client with an operating manual to be based on the manufacturer's data. This is to give a general system description with information on the components, areas and all of the parameters that have been programmed.
- Ensure that the manual describes what effects a fault or power failure can have and any alarm signal warnings that must be investigated. The method for callouts should be included in the manual. Telephone numbers must be included for emergencies.
- Confirm that telephone numbers are included in the manual to cover any general requirements.
- Provide a full demonstration of the system. This is of vital importance as it instils confidence in the client and shows all of the functions that are available. The installer is advised to give a small amount of technical detail but not enough to confuse the client. If the installer does not demonstrate all functions of the equipment the client may lose confidence and feel that the installer lacks knowledge. At the end of the demonstration the engineer is to ensure that the client is confident to use the system in full and is satisfied as to its capability.
- Instruct the client to put a method in hand to advise the engineer of any building works that are conducted in the future so the developments cannot affect the system's integrity.
- In talking to the client it is essential to avoid the use of jargon, never talk down to the customer and provide summaries of written material.
- Be tolerant and careful in listening to the client to understand their concerns and queries. The client will want to add and remove tokens and credentials, obtain new tokens, operate the software, back up logs and deal with visitors and contract staff. These are routine practices and must be understood by the user.

- Confirm the integration with any other security systems or building services and how they interact.
- Refer the client to any manufacturers' data that covers user maintenance such as the cleaning of reading equipment and keypad keys.

Once the demonstration of the entire system has been performed to the client's satisfaction the records can be formulated for future reference. The criteria are:

- System. Details the specification, drawings provided and records how the maintenance and instruction manual is prepared. Notes should be included of any factors that can be used for future reference and any particular features of the subject installation.
- Historical. Log the date and time of system completion. A record of any new visit to the site must be kept with the exact time. Names of the visiting engineers or anyone calling on behalf of the installation company should be recorded. The record is to have a signature of the person making the visit and the work performed. The nature of any fault must be logged with a diagnosis, action taken and the time noted that the work was included. A reason for the visit is to be provided.
- Temporary repair or disconnection. Reasons for any temporary repair or disconnection must be made. Dates are to be logged plus times of replacements of components.

Documentation must be prepared and as a minimum should include the following information. This document may be provided by means of a diagram of the installed system and should be kept in a restricted place:

General

(a) The name, address and telephone number of the controlled premises.
(b) The name, address and telephone number of the customer.
(c) The location and classification of each access point and the type and location of each controller and its associated hardware.
(d) The type and location of power supplies.
(e) Details of those access points which the customer has the facility to isolate.
(f) The type and location of any warning device.
(g) Details and settings of any preset or adjustable controls incorporated into the system.
(h) Any documentation relating to equipment.
(i) The number of keys, codes tokens, etc. to the system provided to the client.

Commissioning data. Confirm:
(a) Correct termination of wiring.
(b) Voltage and resistance at all appropriate points of the system.
(c) Correct alignment and operation of access point hardware and of release and closure mechanisms at each access point.
(d) Correct operation of each reader.
(e) Release time for each door.
(f) Door held open signal, if specified.
(g) Verification of access levels.
(h) Function of system when mains disconnected.

Following the completion of the handover, which should include the signing of contracts by both parties, the installer can confirm the need for maintenance and service and the schedule that this is to follow.

All electronic access control systems are sensitive and need comprehensive and regular maintenance and service as covered in Chapter 7.

6.6 Health and safety

This forms an essential part of the engineer's responsibilities and is indeed a huge subject so can only be covered here in a very condensed form.

The very first health and safety legislation in the UK was passed in 1802. This was clarified in 1974 with the Health and Safety at Work Act (HASAWA), which covers everyone at work, whatever the workplace, with the exception of staff in domestic premises. The Act places clear duties on everyone: employer, employee, the self-employed, manufacturers, suppliers and installers. The main purpose of the Act is to promote good standards of health and safety so preventing people coming to any harm at work. It makes health and safety an essential part of work and it is not an option as it places statutory duties on employers and employees to build safe practices into their work.

Details can be found in regulations made under the Act and examples are the Electricity at Work Regulations1989 and the First Aid Regulations 1981.

The Act is proactive, not reactive, and the framework that it sets up allows for the ongoing process of developing health and safety legislation by updating older regulations and issuing new regulations, some of which originate from the EC.

Most of the responsibility for health and safety falls on employers. They must ensure the health and safety and the welfare of their staff by:

- Providing safe systems of work, safe environments and premises with adequate facilities.
- Providing safe access and egress to and from the workplace.
- Providing appropriate training and supervision.
- Providing information to the employees.
- Having a written health and safety policy if there are five or more employees.
- Providing safe plant, machinery, equipment and appliances with safe methods of handling, storing and transporting materials.

Employers must also ensure that their activities do not endanger people who visit the workplace or members of the workplace or members of the public.

Employees have a duty to take care of themselves as anyone else who may be affected by what they do at work. They must also co-operate with the employer on health and safety matters by following rules and procedures.

The self-employed are covered much in the same way with a duty not to endanger themselves or others by their work activities. The installer working in a private house is governed by the same rules that apply in industry. There is a need therefore to:

- Design and construct a safe product.
- Test the product for safety.
- Provide information and instructions for the user.
- Ensure the safe installation of the product.

An employer is committed to ensure that persons are properly supervised at work, are trained correctly and understand safety procedures. Some human factors that are attributed to causing accidents are:

- Carelessness.
- Inexperience.
- Lack of training.
- Haste.
- Distraction.
- Complacency.
- Breaking safety rules.
- Influence of drugs or alcohol.

In the first instance we shall look at the connection of the access control system to the mains electrical supply.

Connection of the equipment to the mains source

In Section 6.1 we covered the technical requirements for the inspection and testing of the mains supply.

This is governed by the Electricity at Work Regulations 1989 under the Health and Safety at Work Act 1974. The Regulations apply to all electrical equipment and although they have little relevance to domestic properties they do cover work carried out by a contractor in the domestic environment.

There are two primary definitions of duty:

- Absolute. The Regulations are to be applied regardless.
- As far as reasonably practical. This must be considered in conjunction with the Health and Safety at Work Act, Section 40, as precautions or actions taken to satisfy the requirements of duty must be good enough. If any incident occurs the contractor is committed to prove that it was not practical or reasonably practical to do more than what was in fact done.

One of the main occupational hazards that exist with respect to working on electronic access control systems is that they are ultimately powered by the mains electrical supply and this may be at a number of points. In addition in the commercial and industrial environments' multiphase electrical supplies will be encountered. Any person making a connection with the mains electrical source must be competent under the terms of the Health and Safety at Work Act and this must be demonstrated by acquisition of a recognized qualification, or experience or knowledge, that he or she can work safely with electrical supplies. A competent person also recognizes the limits of their expertise and will not undertake work that they are not trained for. The engineer should:

- Never work on a system connected to the mains supply unless recognized as competent under the Electricity at Work Act 1989.
- Never remove labels from plant or equipment that they do not have responsibility for.
- Never assume that circuits are dead.
- Not replace fuses or protection devices and carriers unless they are specifically identified and the appropriate circuit is to be commissioned.
- Never leave fuses or protection devices and carriers in close proximity to a distribution board even if labelling has been carried out.
- Never remove earth wiring, earth tags and identification or remove earthing from metal enclosures or conduit systems and tubing.

At many sites there will exist special requirements related to health and safety and these should be confirmed with the management and safety officers. However, there are measures that exist across the spectrum.

Power tools

These will be used to some extent in every installation. In the main they are electrically driven so basic safety precautions are to be followed in order to reduce the risk of fire, electric shock and personal injury. Safety instructions should be provided with all power tools but the following can act as a guide.

- Check the condition of the plug and cable before use and renew if defective. Ensure that the tool is turned off before inserting the plug in the socket. Ensure that the power supply voltage corresponds with the rating quoted on the equipment identification plate.
- Direct the power cable to the rear and away from the working area such as the point of drilling.
- Apply caution with long hair and loose fitting clothing and jewellery as these could become trapped or drawn into tools.
- If inserting screws with power tools use a slow speed and use extra vigilance with long screws because of the danger of slipping.
- Always have a secure footing when working and even more so when working at an elevated position.
- The workpiece must be securely fixed with the surrounding area clear of obstructions.
- Do not expose power tools to the rain or use them in damp or wet locations. Do not use in the presence of flammable liquids or gases as they can produce sparks.
- Keep the working area well lit.
- Guard against electric shock by preventing any body contact with grounded surfaces such as pipes and radiators.
- Do not force tools as they will work more efficiently at the rate at which they were intended.
- Do not use power tools for a purpose for which they were not intended.
- Use safety glasses and a dust mask if the power tool causes dust.
- Do not abuse the power cord or allow it to come in contact with heat, oil or sharp edges.
- Keep the handles of power tools dry, clean and free from oil and grease.
- Ensure that tool bits are sharp and clean.
- Disconnect the power tool when changing accessories and ensure that these accessories are compatible with the tool.

- Pay attention to power cords and their capacity. Ensure that the insulation is undamaged along its entire length. A guard or other part that is damaged should be replaced.
- Noise levels from power tools can exceed 85 dB(A) and ear defenders should be worn in such cases.
- Promote the use of 110 V tools and step-down transformers in the construction environment.

Percussion drills

These are the most used power tool so can be considered separately.

- Sharp drill bits optimize performance whilst minimizing wear on the tool and/or battery. Only high speed steel (HSS) drill bits in perfect condition should be used to drill metal. Twist drills of this type from 3.5 mm to 10 mm can easily be sharpened with a standard accessory drill sharpener.
- Never place loads on the drill such that it ceases to be able to rotate correctly.
- Change the rotational direction only when the drill is not moving.
- Be aware of setting the drill on percussion (hammer) or standard drilling.
- Tungsten carbide drill bits are needed to drill concrete, stone and masonry. The best drilling is achieved by tungsten carbide bits with hexagon shafts. When drilling tiles the tile should be penetrated by the standard drilling technique before changing to percussion.
- If greater precision is needed use a bench stand.
- Percussion drills need little maintenance other than to periodically clean out the ventilation slots on the motor casing.

Ladders

More than half of the accidents involving ladders occur because they have not been securely fixed or placed and most of these accidents happen when the work is of 30 minutes or less in duration. It is apparent that this is down to haste. Therefore:

- Support the foot on a firm level surface. Never place the foot on loose material or on a further material to gain height.
- Do not tie shorter ladders together to obtain the desired height.
- If possible the top of the ladder should be secured with clips, lashings or straps. As an alternative the base can be secured using blocks, cleats, sandbags or stakes embedded in the ground. 'Footing' is not considered effective for ladders longer than 5 m.

- Make regular inspections of the ladders and do not make makeshift repairs. Check in particular for damaged rungs.
- Do not paint wooden ladders as this can hide defects but use clear varnish.
- Beware of making contact between aluminium ladders and live electrical cables.
- Ladders must extend at least 1.05 m above the highest rung being stood on.
- The angle of use is to be 75% to the horizontal and 1 m for every 4 m in height.
- If the ladder is near a doorway it is to face the door which must be locked shut or secured open.
- Never allow more than one person at a time on a ladder.
- Do not see a ladder as the best device for all applications as a secure scaffold tower is far safer.
- Never carry heavy items up a ladder but use a rope or hoist. Carry tools in a bag or belt holster to enable the free use of both hands to secure a firm hold.
- Never use a ladder at the top of a scaffold tower to extend its height.
- Ladders used to give access to a scaffold tower must extend 1.07 m above the work platform.

Stepladders and trestles

- These are not to be used with any degree of side loading.
- Avoid overreaching when standing on stepladders or trestles.
- Use the correct size devices for the height being worked at.

Scaffold towers

- These are safe to work on but need professional installation.
- When using fixed scaffolds these are to be well lit and closed off to unauthorized use.
- Mobile towers must have the wheels locked in position when at the chosen work position and outriggers are to be used when available.

6.7 Discussion points

It is at this stage that we carry out the installation and siting of the main equipment and the ancillary components. However well a survey has been carried out and the system designed, the installation can still encounter difficulties. The engineer must be thorough in the knowledge of the subject to overcome any problems that may be practical or theoretical. Equally the installer would be at fault if unable to create the highest level of customer care at the handover as this could lead to a loss of confidence by the client.

Discussions can centre on problems that the physical side of the installation can bring despite a thorough survey and specification having been carried out.

7 Service and maintenance. Response organizations

All electronic access control systems need comprehensive and regular maintenance and investigations must be carried out to identify potential faults.

Essentially there are two forms of maintenance, namely preventive and corrective, that cover routine servicing and emergency servicing respectively.

In this chapter we turn our attention to the practices that are required to perform the servicing and to ensure the continued operation of the system. Allied to this we overview the response organizations in this area that include the training providers, inspectorate bodies, trade associations and management agencies who all have an interest in the governing maintenance and the upholding of quality practices in the access control industry.

In the first case there are a number of general points that should be understood with respect to maintenance:

(a) It is advisable that maintenance be carried out by the installation company.
(b) The customer is not obliged to have their system maintained by the installation company and any maintenance is a matter of agreement between the parties.
(c) The maintenance company must have the means, including spare parts and documentation, to effectively carry out the maintenance.
(d) Security is important and safe custody of equipment and documentation must be made. Vetting of employees of maintenance companies must be undertaken and an ID carried to include a photograph, signature, company name and date of expiry.
(e) Standard tools shall be carried and specialist equipment made available for particular investigations.
(f) Staffing levels of the maintenance company must be held with respect to:
    - Number of installations to be serviced
    - Complexity of installations
    - Geographical spread of installations
    - Method of calling out service personnel outside of normal office hours

(g) An emergency service is to be available.
(h) The maintenance company must be able to reach the protected premises within an agreed written period unless abnormal circumstances exist.
(i) Readings, measurements and observations must be recorded for comparison with new criteria and values in the event of a despute arising at a later point in time.

7.1 Preventive maintenance

This is the routine servicing of a system carried out on a scheduled basis.

Within this we include the historical data plus the detail of any work performed and any faults that have been found. This to include operation by the client and any faults generated by the client together with faults attributed to the equipment. In the first instance the system and all of the individual components have been designed with a specific life span or expectancy. Thus, the mean time for the expected failure of certain components, plus a safety margin, should have been taken into account. These factors influence the overall system price and it is from this that the consequential maintenance frequency and costs arise.

Preventive maintenance visits to the protected premises shall be made by a representative of the maintenance company during or before the twelfth calendar month following the month of commissioning or of the previous preventive maintenance visit. The mechanical components in an access control system such as locks and hinges will require routine preventive maintenance by the user more frequently than once per year and this should be brought to the attention of the client.

Even with the best systems regular maintenance is needed. This is because of the demands put on the system by the users, and this is more apparent if it operates in a difficult environment. Preventive maintenance and service includes both testing and inspection but can only be truly effective if the installation company is both trustworthy, competent and the engineers are trained correctly.

This form of servicing is a thorough check in order to prove that all components are working as intended in accordance with the specification and shall continue to do so in the same fashion as when originally installed and commissioned. The inspection is to verify that there are no changes in the environment or area that can adversely affect the system performance. As a minimum the tasks can include:

- System installation
- System use
- System components

System installation

Checks should be made to cover the installation and location of all equipment and components against the original specification and system record. The satisfactory operation of all components should be made.

- Confirm that there are no changes in the environment that could influence the life of any of the system parts.
- Ensure that there is no undue wear attributed to any of the system components that will cause premature failure or rapid deterioration.
- Observe any deviations from the system record or any unauthorized modifications that may have been made.
- Record any observations on the system record.

System use

- Verification of the system use being unchanged from the original specification should be performed.
- The engineer should confirm the building or area has not been developed in any way that can now compromise the system.
- Ensure that the perimeter protection devices remain secure and that no additional access routes exist.

System components

A full check of all of the components and equipment must be performed. They can be considered as groups:

- Perimeter protection.
- Tokens and readers.
- Controllers and power supplies.
- Cabling.

Perimeter protection

This is to include all locking devices and hardware associated with the barriers and doors plus ancillary equipment such as door closers, sensors and push-to-egress buttons.

In practice this is where the majority of faults occur because of the moving parts involved in their operation. Alignment should be checked and any parts requiring lubrication should be serviced.

Locks must not show undue wear and the operating voltage should be verified plus door open and closed timers.

Tokens and readers

Reader heads and keypads are to be cleaned and serviced. Readers could then be checked for the validating of authorized tokens correctly and the rejection of others. Keys on keypads must all function properly.

Non-contact readers must have the range proved and batteries fitted to active tags if appropriate.

Voltages supplied to readers should be confirmed as satisfactory and the results of the validating and voiding by the readers must be verified as held in the log records.

Controllers and power supplies

Power supplies may be combined with a control unit, be separately housed or be British Telecom approved for remote signalling duties. Although they are all essentially reliable they are not serviceable but must be confirmed as not having deteriorated or their values moved out of tolerance. Take readings for the output voltage and current loadings and prove the mains supply and its physical route have not been modified since the original installation. For power supplies with back-up batteries these should be proved to support the system if the mains is disconnected. The charging voltage is to be checked and the batteries changed if any weakness is found and for sealed lead–acid batteries this is to be a maximum of five years. It is difficult to generalize on the function of controllers but they should be checked for off-line or any degraded mode of operation and the log viewed for any working problems during the life of the system. Automatic functions that are not normally used and may have been rarely employed in the system life should be proved.

Cabling

In real terms faults are attibuted more to terminations and jointing or flexible connections and door loops than to the cable.

Circuit measurements can be taken and compared against those logged at the time of the original installation and those taken at subsequent visits.

Containments are to be checked for unauthorized modifications and screens/braids for integrity. If any suspicions of damage to any cables exist the insulation resistance can be checked against the original data.

NACOSS Code of Practice NACP 30 which covers the planning, installation and maintenance of access control systems for its preventive maintenance inspection specifically invokes inspection of the following across the range of systems:

(a) The installation, location and siting of all equipment and devices against the system record. This record is to be generated at the point

of installation and may include previous information from the system design specification. These records are to be protected from unauthorized access.
(b) The satisfactory operation of all equipment.
(c) All flexible connections.
(d) The normal and standby power supplies for correct functioning.
(e) The control equipment in accordance with the company procedure.
(f) The operation of any warning device in the system.

For those items of inspection and rectification which are not carried out during the preventive maintenance visit they are to be completed within a period of 21 days. Those parts of a system or any environmental conditions which are found during preventive maintenance to be the potential cause of reduced security shall be identified on the visit record.

We noted earlier that certain mechanical components need more regular maintenance than many of the system components. This hardware, such as locks and hinges plus release mechanisms, will originally have been selected in accordance with the degree of security related to the classification and anticipated traffic and duty cycle. This equipment must therefore be seen in its own right.

At the visit the service engineer should assess the areas that such hardware is operating in to ensure that the following are not creating problems:

- Temperature.
- Humidity.
- Corrosion.
- Vibration.
- Dust and other contamination.
- Physical abuse.

At the conclusion of the visit the results of the inspection are to be entered on a maintenance visit record along with the signature of the client or representative. A copy should be given to the client.

A historical record with the date of every visit, faults found and action taken is also to be kept. Details of every fault reported to the maintenance company are to be included with details of the action taken and if known the cause.

All information shall be kept for at least 24 months after the inspection to which it refers.

7.2 Corrective maintenance

This is the emergency servicing in response to the development of a fault. It can also be as a consequence of a fault found during a preventive maintenance visit.

Successful and efficient fault finding or troubleshooting depends on logical reasoning and good organization. It is vital that the engineer can understand the circuits in which they are interested and how these function. How circuits are actually functioning in relation to how they are actually working is the basic information needed.

The electronic access control engineer may not specialize in the fundamentals of electronic components but it is of help if he or she is able to interpret circuit and block diagrams and identify the functions of the components in a circuit as this will make fault finding more logical.

With many systems it is possible that the engineer can be guided by the control log or displays on indicating equipment as these may establish the point at fault and the solution may be a simple exchange of components or ancillary devices. Certainly some electronic devices do fail. All equipment and many ancillary devices incorporate semi-conductors, diodes, transistors and integrated circuits which in principle are reliable yet are manufactured in volume. These components when given their rated value are also attributed with a tolerance. Accepting wide tolerances means that the unit cost can be lower but the designer must ensure that no matter what combination of tolerances exists, the final product must still function as intended. Therefore the designer is committed to design not only for economic manufacture but also for future maintainability and service. Devices must also be constantly developed to retain their market potential and changes must be implemented and assessed quickly so that improvements in reliability are gained. However, components continue to fail as technology advances and this, allied to the unfeasibility of checking every component means that uncertainties will always exist. However, the evidence is that electronic equipment is essentially inherently reliable. The installer in the first instance will use experience in the selection of equipment so that the installation can stand the test of time.

Mechanical hardware faults may be more easily traced and rectified. An access control system can also be broken down into sections to enable a rapid fault detection if electronic devices are clearly troublesome. However, some of the faults attributed to cabling are difficult to detect and it is possible that at the design stage certain fundamentals were overlooked and problems then manifest later on.

Major cabling types are always recommended by the manufacturer for the principal equipment and considerations for design are covered in Sections 5.2 and 6.3 for connection and testing. It must therefore be clearly understood that a badly conducted survey and system design will result in a system that will be troubled by ongoing problems. Badly engineered goods and systems cannot be readily cured by corrective maintenance.

However, it still remains that problems occur even with well-designed systems and specifying good quality equipment. Nevertheless there are time-honoured steps that can be taken to establish the perspective of dealing with problems that occur with any system and these can be related to the access control industry even though there is a proliferation of equipment in use:

- Understand the characteristics of the equipment and system that are being worked on.
- Understand how the equipment is supposed to operate.
- Research the symptoms that give indication of incorrect operation.
- Determine the form of failure that is the cause of these symptoms.
- Isolate the problem by breaking the system down into parts and eliminating the non-troublesome parts.
- Rectify the fault.
- Verify that the system once again functions as intended and prove that the correct measures have been taken and that no further problems are in existence.

Experience will show that once a fault has been traced and corrected it should not be assumed that the full system problem has been overcome. It is possible that the system fault could be an accumulation of small problems and that the repair only corrected a portion of the system trouble. Equally, one of the faults may be time related and not be existent at the time of the corrective maintenance. These intermittent and accumulated problems can be the most difficult to cure.

A learning curve can be formed with many goods and the installer may want, within reasonable limits, to restrict the use of specified equipment to a small and easily identified number of manufacturers with whom he or she is familiar. This enables the installer and service personnel to become more aware of potential problems attributed to particular goods.

Unfortunately progression will lead the installer to diversify so a medium must be reached. It cannot be possible for the engineer to become settled in a particular niche and also tender for the future.

In addition the user should be instructed to hold a log of any claimed faults and the source/person from which they have been reported. This is important because some faults may be a clear case of operator error.

The installation company must also hold a database on corrective maintenance, the remedial action taken and the outcome.

The corrective maintenance record must include the date and time of receipt of each request for emergency service, together with the date and time of completion of corrective maintenance and the action carried out.

The results of a corrective maintenance inspection are to be entered on a maintenance visit record and the signature of the client or his or her representative obtained on the record. A copy of the record should be given to the client.

All information should be kept for 24 months after the inspection to which it refers. If a preventive maintenance inspection is made at the same time as a corrective maintenance visit, separate records should be completed.

There will be times when it is not possible to cure a fault immediately. In the event that a temporary disconnection must be made this must be recorded. This must identify the parts of the system that are not operable. The reason for the disconnection, the date and time of disconnection and of subsequent reconnection must be given. A signed authorization for each disconnection should be obtained from the customer.

This authorization should be kept for three months after reconnection.

7.3 Response organizations

In response to the availability and growth of the various electronic security systems, including the electronic access control area, many active bodies throughout the world have developed to manage the industry and promote high quality practices.

Organizations

The British Security Industry Association (BSIA)

This body acts in the industry to ensure that all of its members meet stringent requirements, work to BSIA codes of practice and comply with relevant British Standards for their products. This gives a guarantee of a secure product or service to any customer choosing a BSIA member company. It is the most effective security grouping in the UK. It also produces technical literature, guidance notes, codes of practice and training materials and is the preferred source of information for the media on the security industry.

Membership sections extend from access control through to the other security techniques.

Security Industry Training Organization (SITO)

Created by the BSIA to manage training to the long-term benefit of the security industry, SITO is the principal training provider of all electronic security concepts. It extends its interests into other security areas including physical protection and guarding.

Loss Prevention Council (LPC)

A testing organization with interests in security risks.

Association of British Insurers (ABI)

Overviews the security industry with a view to assessing risks.

Inspectorates

National Approvals Council for Security Systems (NACOSS)

The industry governing body and the most influential of all the inspectorates. NACOSS carry out inspections on all equipment installed by their member companies to endorse their compliance with the related standards and NACOSS Codes of Practice. A self-regulatory body, they hold a list of recognized firms in order to give assurance as to quality of installation and maintenance. Any recognized firm must have a QA management system to BS EN ISO 9002 plus NACOSS's own quality schedule.

Security Systems and Alarms Inspection Board (SSAIB)

Operates essentially in the same fashion as NACOSS in carrying out technical and company inspections on installers. The Electrical Contractors Association (ECA) is a member.

Alarms Inspectorate and Security Council (AISC)

A further inspectorate with approved companies ranging from national installation companies to local government bodies and independent small businesses.

Integrity 2000

Has a number of categories for installers of security equipment and is prominent in access control to oversee installers wanting credentials of third party assessment.

Standards producing bodies and approvals authorities

British Standards Institution (BSI)

BSI is responsible for the compilation of British Standards to govern the technical practices of any particular industry. These standards are the consensus of opinion of the various interested parties to endorse good working practice and consistency of manufactured products.

The British Standards Institution through its working parties and standards writing committees liaises with the principal international standards producing bodies:

- International Electrotechnical Commission (IEC).
- International Standards Organization (ISO).
- European Committee for Electrotechnical Standardization (CENELEC).

Underwriters Laboratories (UL)

Have a major interest in North America with worldwide recognition.

An independent non-profit making corporation founded to test products for public safety, with their services used by the insurance industry to assess risks. Approval by UL, in that third party assessment of goods must be made, is mandatory in many parts of the USA for non-military electrical goods.

Canadian Standards Association (CSA)

Operate in a similar fashion to UL and tend to endorse similar standards. Have an influence on standards and endorsement of good working practices.

In North America and other parts of the world that are guided by American practices there are other organizations that have an influence on standards and the endorsement of good working practices:

- American National Standards Institute (ANSI).
- American Society of Industrial Security (ASIS).
- American Society for Testing Materials (ASTM).
- Electronic Industries Association (EIA).
- National Burglar and Fire Alarm Association (NBFAA).
- National Electrical Manufacturers Association (NEMA).
- National Institute of Justice (NIJ).
- Security Industry Association (SIA).

The private security industry will always be seen as a diverse industry that strives to reduce crime for businesses and the general public. For this reason this sector will always be asked to maintain and improve standards as a whole and licensing schemes are being considered in the UK to cover those who work with security systems in an effort to regulate the industry.

Licensing is intended to vet and remove those with criminal records and the Private Security Industry Authority (PSIA) may be instructed to implement such schemes as a future policy. This organization will work alongside the Association of Chief Police Officers (ACPO) and the UK Accreditation Service (UKAS). The latter organization is a government controlled body which works as an industry inspectorate effectively carrying out checks on the response organizations themselves.

It remains to say how fluid the security industry must remain and how the future will surely unfold.

7.4 Discussion points

The client accepts that there is a clear need for corrective maintenance as a problem area must be corrected; however, some customers are reluctant to enter into a preventive maintenance contract and only want service calls following a breakdown of the system. For this reason discussions must clearly identify the purpose and logic of preventive maintenance schedules.

The importance of identifying the requirements of regulatory bodies, standards and codes of practice to ensure that the system conforms to accepted criteria and methodology should be addressed.

The system and why it must be seen as whole that the failure of one part then constitutes a deviation from the design specification are further topics for discussion.

8 Reference information

8.1 Environmental protection

All components for both indoor and outside use are governed by their ability to withstand extremes of climate, changing ambient temperature limits, the weather and the ingress of liquids and dusts. The degreee of protection is indicated by the letters IP followed by two characteristic numerals. Reference can be made to standards BS 5420, IEC 144 and IEC 529.

The first numeral indicates the protection afforded against the ingress of solid foreign bodies and the second the protection against the ingress of liquids.

It should be appreciated that many goods are supplied in housings and are coded to indicate the classification when gaskets are correctly seated and conduits, etc. sealed.

If the letter W is used as a supplement, it indicates particular features for use under specific weather conditions and as endorsed in manufacturers' data.

First characteristic numeral. Degree of protection. Solid bodies	*Second characteristic numeral. Degree of protection Liquids*
0 No protection	0 No protection
1 Protection against ingress of solid bodies larger than 50 mm, e.g. a hand.	1 Protection against drops of condensation.
2 Protection against ingress of medium size solid bodies larger than 12 mm, e.g. a finger.	2 Protection against drops of liquid falling at an angle up to 15° from the vertical.
3 Protection against ingress of solid bodies greater than 2.5 mm thick.	3 Protection against rain falling at an angle up to 60° from the vertical.
4 Protection of solid bodies greater than 1 mm thick.	4 Protection against splashing water from any direction.
5 Protection against harmful deposits of dust interfering with the satisfactory operation of the equipment.	5 Protection against jets of water from any direction.
6 Complete protection against dust.	6 Protection against water similar to that encountered on ship decks.
	7 Protection against immersion in water specified pressure and time.
	8 Protection against prolonged immersion in water.

North American practices and those countries that are influenced by American standards use the NEMA system with the classifications contained in publication No. ICS-6. In general those that we are interested in are:

Type 1. General purpose, indoor.
Intended to prevent accidental contact of personnel in areas where unusual service conditions exist. Protection is provided against falling dirt. The enclosures may or may not be ventilated.
Type 2. Drip proof, indoor.
Intended to protect the enclosed equipment against falling non-corrosive liquids and falling dirt. These enclosures have provision for drainage.
Type 4. Watertight and dust tight, indoor and outdoor.
Intended for use to afford protection against splashing or hose directed water, seepage of water or severe external condensation. They correspond generally with IP54.
Type 6. Submersible, watertight, dust tight and sleet resistant, indoor and outdoor.
Intended where occasional submersion is encountered. This classification is generally as IP67.
Type 13. Oil tight and dust tight, indoor.
Intended to afford protection against lint and dust, seepage, external condensation and the spraying of oil, water and coolant. This classification is generally as IP65.

8.2 Multiplication factors

Factor	*Prefix*
1 000 000 000	Giga (G)
1 000 000	Mega (M)
1000	Kilo (k)
100	Hecto (h)
10	Deca (da)
0.1	Deci (d)
0.01	Centi (c)
0.001	Milli (m)
0.000 001	Macro (μ)
0.000 000 001	Nano (n)
0.000 000 000 001	Pico (p)

8.3 Common multiples

Unit	*Multiple*	*Value*
Ampere	Milliampere (mA)	1/1000 ampere
Ampere	Microampere (μA)	1/1 000 000 ampere
Volt	Millivolt (mV)	1/1000 volt
Volt	Microvolt (μV)	1/1 000 000 volt
Ohm	Kilohm (kΩ)	1000 Ω
Ohm	Megohm (MΩ)	1 000 000 Ω

8.4 Standards. Codes of practice

In this section we note the documents that can be referred to with relation to access control systems. There are many instances that these do not directly relate to the specific subject of electronic access but in these cases they can be used as a guide. In other cases they form part of integrated techniques.

Access control and locks

British Standards (BS)

BS 2088: Performance tests for locks.
BS 3621: Specification for resistant locks.
BS 5872: Locks and latches for doors in buildings.

BS/EN European

BS EN 50133: Alarm systems. Access control systems for use in security applications.
BS EN 50133-1: System requirements.

British Security Industry Association (BSIA)

BSIA 107: Access control systems. Planning, installation and maintenance.

NACOSS Codes of Practice

NACP 1: Code for security screening of personnel. BS 7858.
NACP 2: Code for customer communications.
NACP 3: Code for management for subcontracting.
NACP 4: Code on compilation of control manual.

NACP 5: Code for management of customer complaints.
NACP 30: Code for planning, installation and maintenance of access control systems.

Underwriters Laboratories (UL)

UL 291: Access control units.
UL 768: Combination locks.
UL 887: Delayed action time locks.

American Society for Testing Materials (ASTM)

F12.60: Controlled access, security search and screening.

Health and safety

The Health and Safety at Work Act 1974.
The Factories Act 1971.
The Offices, Shops and Railways Premises Act 1963.

Remote signalling and central stations/alarm receiving centres

British Standards (BS)

BS 6301: Requirements for connection to telecommunication networks.
BS 6305: Requirements for connection to British Telecommunication telephone network.
BS 6320: Modems for connection to telecommunication networks.
BS 6329: Specification for modems for connection to PSTN.
BS 6701: Installation of apparatus intended for connection to certain telecommunication systems.
BS 6789: Part 3: Apparatus for automatic calling and answering.

EN European Standards

EN 50136.1.3: Alarm transmission systems for digital PSTN.
EN 50136.1.4: Alarm transmission systems for voice PSTN.
EN 50136.2.3: Requirements for digicoms.

Underwriters Laboratories (UL)

UL 827: Central station for watchman, fire alarm and supervisory services.

Security Industry Association (SIA)

ART. 725: Remote control signalling circuits.

Quality

Harmonized

BS EN ISO 9000: Quality management in quality assurance standards.
BS EN ISO 9000-1: Guidelines for selection and use.
BS EN ISO 9001: Quality systems. Model for quality assurance in design, development, production, installation and servicing.
BS EN ISO 9002: Quality systems. Model for quality assurance in production, installation and servicing.
BS EN ISO 9003: Quality systems. Model for quality assurance in final inspection and test.
BS EN ISO 9004-1: Guidelines.

Reference

British Standards (BS)

BS 3116. Part 4: Specification for automatic fire alarm systems in buildings. Control and indication equipment.
BS 3621: Thief resistant locks.
BS 4737. Part 1: Intruder alarm systems in buildings. Specification for installed systems with local audible and/or remote signalling.
BS 5420: Degrees of protection of enclosures for low voltage switchgear and controlgear.
BS 5839: Fire detection and alarm systems in buildings.
BS 5872: Locks and latches for doors in buildings.
BS 6360: Conductors in insulated conductors and cables.
BS 7042: Specification for high security intruder alarm systems in buildings.
BS 7671: Requirements for electrical installations. Issued by the Institute of Electrical Engineers as the IEE Wiring Regulations.
BS 7807: Integration of fire and security systems.
BS 8220: Guide for security of buildings against crime.

British Security Industry Association (BSIA)

BSIA 195: EMC Guidelines for installers of security systems.
BSIA 256: Guide for Paknet radio alarms.
BSIA 284: Guide for the use of downloading in the security industry.

Underwriters Laboratories (UL)

UL 609: Burglar alarm systems – local.
UL 611: Burglar alarm systems – central station.

Index